PRAISE FOR *SAVED BY SCIENCE*

"Over the next few decades, we will learn how to edit life-forms by altering their DNA, much as we now edit stories and books using ABCs. Poznansky gives us a sense of what this might mean for medicine, food, climate, and a myriad of human endeavors and challenges. He does so with grace and ease, such that any curious mind can comprehend." — JUAN ENRIQUEZ, author of *As the Future Catches You* and co-author *Evolving Ourselves.*

"*Saved by Science* is a terrific gateway into the fascinating world of synthetic biology. It outlines why we need these new genetic superpowers more than ever if we are to solve global challenges and continue our most exciting explorations, such as establishing thriving colonies on Mars. Mark Poznansky's book is a tonic for anyone concerned about global hunger, energy, health, or climate. It turns out we've got the tools to address pretty much every challenge right at our fingertips — we just have to get to work." — ANDREW HESSEL, founder and president of Humane Genomics

"Synthetic biology promises to be the next important step in our application of DNA. Mark Poznansky not only introduces us to the incredible potential of constructing novel living organisms but also provides the context to ensure that the inevitable debate will be constructive." — JAY INGRAM, science writer and broadcaster

SAVED

_____

BY

SCIENCE

DR. MARK J. POZNANSKY, C.M., O.Ont.

SAVED

───────── BY

SCIENCE

The Hope and Promise of Synthetic Biology

Purchase the print edition
and receive the eBook free.
For details, go to ecwpress.com/eBook.

Published by ECW Press
665 Gerrard Street East
Toronto, Ontario, Canada M4M 1Y2
416-694-3348 / info@ecwpress.com

Editor for the Press: Jennifer Smith
Cover design: David Drummond

LIBRARY AND ARCHIVES CANADA CATALOGUING IN PUBLICATION

Title: Saved by science : the hope and promise of synthetic
biology / Dr. Mark J. Poznansky, PhD.

Names: Poznansky, Mark J., author.

Description: Includes bibliographical references and index.

Identifiers: Canadiana (print) 20200235931
Canadiana (ebook) 20200236067

ISBN 978-1-77041-535-5 (hardcover)
ISBN 978-1-77041-603-1 (paperback)
ISBN 978-1-77305-595-4 (PDF)
ISBN 978-1-77305-594-7 (EPUB)

Subjects: LCSH: Synthetic biology. | LCSH: Science—
Forecasting.

Classification: LCC TA164 .P69 2020 | DDC 660.6—dc23

The publication of *Saved By Science* is funded in part by the Government of Canada. *Ce livre est financé en partie par le
gouvernement du Canada.* We also acknowledge the contribution of the Government of Ontario through the Ontario Book
Publishing Tax Credit, and through Ontario Creates for the marketing of this book.

PRINTED AND BOUND IN CANADA

PRINTING: FRIESENS   5   4   3   2   1

This book is dedicated to my wife, Ilona Feldman, who puts up with a lot; to my daughters, Shoshana and Mirit; and to all of our families. But most of all, it is dedicated to our grand-children: Avi, Ephi, Olivia, Gili, Zachary and Rafi (as of this writing), who are inheriting a world that is far from pristine. They will need all these technologies and more to prosper and live at the standards that many of us have enjoyed.

Table of Contents

Preface

The future of mankind is far from secure. I am among many who believe that humanity is in crisis; in particular, our personal health, the security of our food supply and the health of our environment all face potentially catastrophic challenges. Our health faces many unresolved dangers in the areas of cancer, infectious diseases and mental health. Rapid population growth and the many environmental challenges in our agricultural systems raise questions about how we will feed the world in the year 2050. Global warming and climate change are threatening our environments, and pollution is poisoning our land, lakes, rivers and oceans.

While these challenges are monumental and the future may appear bleak, there is hope. Imagine being able to:

- Identify specific genetic mutations of a whole range of cancers and to develop personalized and specific therapies (i.e., cures), even at the patient's bedside.
- Modify the genetic mutation that predisposes people to suffer from schizophrenia, bipolar disease, severe depression or addictive disorders and to offer effective cures.

- Respond to any viral outbreak (such as Ebola, Zika, AIDS, a nasty flu or COVID-19) with an effective vaccine produced in only days or even hours.
- Grow nutritious, inexpensive, high-protein foods in the widest range of possible conditions of temperature, sunlight, water and fertility . . . or even on Mars.
- Create real meat without killing animals or to produce real milk without milking cows.
- Provide plants with nitrogen from the air instead of having to mine or chemically synthesize expensive nitrogen fertilizers.
- Reverse global warming by removing carbon from the atmosphere and using it as an energy source or material for advanced manufacturing.
- Use microbes to clean up lakes and rivers, removing lead, mercury and other toxic materials and returning our waterways to pristine condition.
- Design specific microbes to clean up toxic-waste dumps, abandoned mines and industrial sites, and even to clean up disastrous oil spills.

A mere six or seven years ago, these imaginings would have been purely the stuff of science fiction. Today, we have realistic expectations that they'll happen — and that they'll be brought to market within a decade, maybe even less. These are the products of what some call the "fourth industrial revolution," a marriage of computer science and newfound knowledge in biology, particularly genomics. This book is about that revolution, a new field of science called synthetic biology and the hope and promise that it offers for the future of mankind.

A note to readers: This book is aimed not at a science audience but, rather, at people who want or need to know how humans are going to overcome some of our major hurdles in health, food and the environment. Following the main text, I've included a glossary and a genetics primer that may help you navigate some of the science that, by necessity, is included in the book.

Chapter One

INTERSECTION

This is a book about change and its many ramifications. In the past half century, social and physical changes such as population growth and industrialization have happened rapidly and had serious consequences. In fact, the pessimistic part of my brain has some pretty grave worries about the future of humanity because we're facing some especially difficult challenges in areas that I refer to as the "big three": health, food and the environment. How are we going to cure the diseases that so many still suffer from? How are we going to feed an ever-increasing population? How are we going to solve our many environmental problems, including global warming? These are big problems that seem to be thrown at us with ever-increasing frequency and severity.

Fortunately, however, change is not limited to these many grave issues. There are also many promising changes happening, particularly in the areas of computer science and biology (especially genomics), and so, the optimistic part of my brain tells me to chill. It tells me to look at the advances we've made in technology and science over the past 50 years and to have faith that real solutions will come from research and development and that they'll come in time. This book is born of that faith and, in particular, of my belief in the promise of a new science — synthetic biology — and its potential to solve some of humanity's most serious problems.

1

Shortly before his death, Steve Jobs, of Apple fame, said:

**"I think the biggest innovations of the
21st century will be at the intersection of
biology and technology."**

I agree. Synthetic biology is built at that intersection and focuses on our expanding knowledge of biology and genomics and our ability to engineer new biological systems or applications (apps). Ontario Genomics, a nonprofit government-funded agency in Canada, defines synthetic biology as "the science of building simple organisms or 'biological apps' to make manufacturing greener, energy production more sustainable, agriculture more robust and medicine more powerful and precise."

When I use the term "biological apps," I'm referring to new life-forms, whether completely novel or partly modified. Examples currently being developed include:

- a novel virus whose sole function is to search out and destroy cancer cells with particular genetic mutations;
- a genetically modified white blood cell (perhaps your own) whose sole function is to search out and destroy cancerous tissue;
- a new form of lettuce that contains a significant amount of protein and can grow with 50 percent less sunlight and 75 percent less water than traditional lettuce;
- a new bacterium that will take carbon dioxide out of the atmosphere and reuse it to make building materials for housing, resulting in a decrease in greenhouse gases and reversal of global warming; and
- a microbe genetically modified to help remove mercury, lead and other pollutants from our lakes and rivers.

It's this science that gives me faith that we'll be able to generate important solutions to our existing problems, including the big three, and position ourselves to face the many new threats to humanity that are on the horizon.

In the chapters that follow, I'm going to introduce some innovative ideas and technologies coming out of the digital revolution and the explosion of knowledge in genomics and biology. I'm also going to outline some of the rather critical stresses that humanity faces in the first part of the 21st century. I'll describe the challenges in some detail because, in our comfortable Western, often middle-class, world, we may feel protected or immune from a multitude of threats, and this provides us with a dangerous and false sense of security. And, because I'm an optimist, I'll offer exciting solutions based on the new science of synthetic biology.

THE BIG THREE: HEALTH, FOOD AND THE ENVIRONMENT

I'm a pretty healthy 70-plus-year-old, so I have few personal health concerns, but broadly speaking, there are many medical and health issues that afflict people but are not yet being successfully addressed. For example, what would happen if there were a local outbreak of the Ebola virus in a major population center in North America? As I write this, there is not yet an effective vaccine or treatment for the COVID-19 virus, and that is just one of a host of potentially lethal infectious diseases that could wreak havoc in populated areas. In addition, many cancers remain unresolved. For instance, there has been very little progress toward a cure for pancreatic cancer in spite of many hundreds of millions of research dollars aimed at seeking a cure. And while we've made moderate progress in addressing some of the stigmas of mental illness in recent years, the truth is that over the past 45-plus years, there has been frighteningly little progress in the treatment of severe psychosis (such as bipolar disease, depression and schizophrenia) or many other, often devastating, mental illnesses, including addiction.

While it's natural for us to be most concerned about our own mortality (read, health) and that of our friends and family, humanity also faces other serious problems: the security of our food supply and the health of our environment. The fact is that if you live in the Western world in 2019, you'll seldom, if ever, notice food shortages. Food is plentiful and relatively cheap, and it's been calculated that we throw out at least as much food as we consume. But globally, more than 12 million people starved to death in 2018 (including more than three million children), and over a billion suffer from moderate to severe malnutrition. The death rates (per 100,000 population) from malnutrition vary dramatically across the world, from 121 in Central Africa to 12.5 in Bangladesh, 2.25 in France and 0.11 in Germany. The whys and hows are of critical importance: Is this a function of extreme poverty or is it an issue of the distribution of food to impoverished regions and areas where food production is limited? I will address that question plus the issues of growth in population and the urban classes, changes in food production and whether we are heading toward serious food shortages, such as the large global food gap that many experts say we'll experience by the year 2050, unless some major changes take place.

Finally, unless you've been living in a cocoon or have been on an interplanetary voyage for many years, you must have noticed that our climate has changed dramatically. Our winters are warmer, as are our summers — and our oceans. The polar ice caps are melting, glaciers are retreating, sea levels are rising and weather patterns are changing in such a way that we increasingly experience severe weather, droughts and storms. I'll discuss what the future might hold for climate change and what we might do to counter this serious threat.

SYNTHETIC SOLUTIONS

Information technologies have been advancing at a tremendous rate, which has enabled us to compile and interrogate billions (if not trillions) of bits of information — amounts of data that were unfathomable just a short time ago. Meaningful advances are also being made in biology, especially genomics, which is our ability to read and understand the code of life — not only our own (the human genome) but also potentially that of every living organism, including microorganisms.

The combination of advances in both of these fields has given rise to the relatively new field of synthetic biology (sometimes also known as synthetic genomics), defined by the Engineering Biology Research Consortium as: "The design and construction of new biological entities such as enzymes, genetic circuits and cells or the redesign of existing cells." While in its nascent stages, synthetic biology is not entirely new. In most cases, we're simply going to take lessons from billions of years of microbial evolution and use them to help mankind adapt to the many stresses of industrialization and a rapidly increasing population.

I'm going to show you how synthetic biology is taking these lessons and using them to introduce new technologies that help us to better understand cancer and to detect, attack and kill it with a much more "intelligent" or "biological" approach. Similarly, I'm going to describe novel approaches to developing vaccines against killer viruses such as Ebola and COVID-19, and I'm going to examine new ways to understand and develop treatments for mental illnesses such as depression and schizophrenia.

I'll show you how we'll be able to create better and healthier food to feed Earth's growing populations and how it'll be done using less fertilizer, less energy and less water, thus diminishing the agriculture industry's huge carbon footprint. I'll show you how we can use synthetic biology to help stop global warming — and even reverse it — and start reusing emitted carbon rather than simply trying to store it. I'll describe how we're learning to clean up the environment by using techniques that microbes have been using for billions of years to sequester lead, mercury and even uranium, all major contaminants that to date we have had difficulty dealing with.

These potential solutions are exciting, as is the future of synthetic biology, especially considering the ever-increasing pace of innovation, which is highlighted in this quote from prolific novelist Dan Brown's 2017 novel, *Origin*:

> It took early humans over a *million* years to progress from discovering fire to inventing the wheel. Then it took only a few *thousand* years to invent the printing press. Then it took only a few hundred years to build a telescope. In the centuries that followed, in ever-shortening spans, we bounded from the steam engine, to gas-powered automobiles, to the Space Shuttle! And then, it took only two *decades* to start modifying our own DNA. We now measure scientific progress in *months*. It will not take long before today's fastest supercomputer will look like an abacus, today's most advanced surgical methods will seem barbaric and today's energy sources will seem as quaint to us as using a candle to light a room.

It's an interesting time when problems and solutions seem almost to drive one another. In *New York Times* columnist Thomas Friedman's book *Thank You for Being Late: An Optimist's Guide to Thriving in the Age of Accelerations*, he focuses on the rapid changes going on around us, giving a dizzying account of the technological advances that have occurred over the past two decades. To realize how far we've come, we have only to think of the changes in our home computers: how fast they've become and how much data they can store. Remember *The Encyclopedia Britannica*? Many of us had copies in our homes or maybe in our parents' or grandparents' homes. The last copy was published in 2010 and was made up of 32 volumes, each comprising 1,375,000 words, taking up most of a large bookcase. Today we can store 139 complete copies of the encyclopedia, or 4,200 volumes and around six billion words, on a simple 128 gigabyte USB key and only use about a third of the available space. And what if there was a simple error on one of those printed pages? It could take you years to find it. You probably never would. My iPhone could find the mistake in a matter of seconds, if not less. Those are the types of changes and "accelerations" that Friedman is talking about.

While he's focused on information technologies, changes in our understanding of genomics and biology have also progressed at incredible rates, fueled at least partially by the digital revolution. In one of his book's chapters, titled "Just Too Damned Fast," Friedman refers to a quotation from Ray Kurzweil, an American author, computer scientist, inventor and futurist, who put it beautifully: "We're entering an age of acceleration. The models underlying society at every level, which are largely based on a linear model of change, are going to have to be redefined. Because of the explosive power of exponential growth, the 21st century will be equivalent to 20,000 years of progress at today's rate of progress: organizations have to be able to redefine themselves at a faster and faster pace."

In progressing through this book, think about the incredible changes that have occurred just in our lifetimes, the ever-increasing pace of innovation and Friedman's "accelerations." Think about the new mega companies, Amazon, Google and Facebook, and how they have very quickly changed many aspects of our day-to-day lives. Think about the pace at which serious problems are being presented, but also think about the pace at which novel solutions are and can be realized.

A mere five years ago, had we looked at some of the potential solutions that synthetic biologists are currently developing, this book would have been considered a collection of science fiction, simply exciting dreams for the future. Today, the marriage of the digital and biological revolutions is launching us into a new era, one in which those dreams will become reality. And so, in spite of the pessimistic part of my brain, I am optimistic — I have no doubt that, as we enter this new era that begins at the intersection of synthetic biology and other technologies, we will engineer successful solutions to many of humanity's most serious problems. In reality, we have no choice.

Chapter Two

LESSONS FROM BIOLOGY

Many of the solutions to the world's most problematic issues will owe their existence to the massive increase in knowledge of our own biology and that of many other species, especially the microbes with which we coexist. Microorganisms, or microbes for short, have been around for about 3.5 billion years, play a critical role in our existence and are all around us. Recent estimates suggest that 70 percent of Earth's biomass is made up of vegetative material while, at 29 percent, microbes make up the vast majority of the rest. Animals, both terrestrial and marine, account for less than 1 percent — and humans come in at less than 1/100th of 1 percent. In addition to being all around us, microbes (or bacteria) are also inside us. It's been estimated that every one of us has about 39 trillion microbes living within our bodies; for comparison, each of us has roughly 30 trillion human cells.

We're learning a lot about and from microbes, and they contribute immensely to the new field of synthetic biology. In recent years, the biotechnology industry has learned to exploit the activity of many microbes to our benefit, using them to make medicines and to help in the production of various foods.

GENOMICS

A critical component of this new knowledge in biology (both about ourselves and about microbes) is represented by the modern field of genomics, and, to a large extent, our increasing knowledge is driven by our capacity to generate and interrogate massive amounts of data at speeds never dreamed of just a decade ago — another nod to Jobs' assertion that the 21st century will belong to the intersection of technology and biology.

First, a note about the words "genetics" and "genomics," which are often used incorrectly or even synonymously — which they are not. Genetics is the study of single genes and their roles in the way traits or conditions are passed from one generation to the next, whereas genomics is the study of all parts of a human's (or any other organism's) genes, known as the genome. Someone once suggested a botanical analogy, whereby genetics represents the individual flowers in a garden and genomics refers to the garden in its entirety. It's a subtle difference, but one worth keeping in mind.

Let's for a moment get back to the "big three" problems facing mankind. Most of the major challenges that humans face — overpopulation, disease, food shortages, climate change, insults to the environment — came about as a result of the same pressures. These include population growth, urbanization and industrialization, and each in turn has two general characteristics: (1) for the most part they are man-made and (2) they are an assault on biology. By that I mean they put undue stress on biological systems. For humans, this can involve living in close quarters and having to drink contaminated water; for the African elephant, it can mean being overhunted and consigned to game reserves; for the North Atlantic cod, being overfished and increasingly maintained in constrained fish farms; for the monarch butterfly, having its habitats overtaken by urbanization; for the whale, being exposed to warmer waters and massive increases in boat traffic and for the ABCD (that's any) flower, it can mean being removed from its natural habitat and bred for our pleasure.

In developing solutions to these problems, we will depend heavily on the biology of microbes and microbial evolution. Indeed, at its core, the entire concept of evolution is based on genetics. Let me explain. In his seminal work, *On the Origin of Species*, published in 1859, Charles Darwin laid the groundwork for all that we understand about the processes by which organisms (whether they be microbes, plants, birds or humans) adapt to new conditions or stresses, and so, how they evolve.

Historically, the power of biology lies in *adaptation*. (We don't see it so much in ourselves, largely because we're so young in terms of evolutionary time.) This process is described by Darwin's theory of natural selection, and we now know that adaptation involves changes in DNA — or "mutations," if you like — that are, more often than not, random in nature. If positive, these changes allow organisms to adapt to new living conditions by altering the way they function (or even exist) in response to the increased stress, the mutation having given the organism some sort of "advantage." However, the kicker is that the process of adaptation in nature is not specific — as in humans adapting to cold or noise or even poverty by changing their behaviors — but rather, random, and therefore slow.

Let's look at a simple example. Say that an organism such as a species of flowering plant finds itself increasingly under pressure or stress from some source, perhaps a decrease in temperature. If the temperature gets too cold and if no advantageous adaptations arise in the plants, then they will likely die and, in time, the species might even become extinct. If, however, through a random (chance) mutation or mutations, even one of the plants develops a strategy to protect itself against the cold, then that species, in its newly adapted form (in this case, maybe just a single mutated gene), may survive and flourish. I use the word "strategy" loosely, because the beneficial mutation was likely not an intentional change but a random event that allowed the species to adapt and survive.

So why don't we simply adapt to the changes and stresses I've been talking about in the same way that Darwin theorized species do through natural selection? The simple answer is *time*. It takes time — and sometimes a lot of time — for random mutations and natural selection (that is, Darwinian evolution) to take, and with the stresses being experienced today by many life-forms on Earth, which are accelerating at a rapid rate, we simply don't have the luxury of that much time. Of course, though, it's not just about time. These are random events, so there is also a lot of chance involved, but the main element is time. Just think for a moment about the time frames of evolution or genetic adaptation and, in the absence thereof, extinction.

Figure 1: Time Frame of Evolution

	Years
Industrialization	220
Civilization (as we know it)	6,000
Human evolution	300,000–500,000
Plant evolution	700,000,000
Microbial evolution	3,500,000,000

Given the fact that many, if not most, genetic mutations are indeed random, any specific mutation is more likely to have occurred in a microbe, because there are many more of them than there are humans, and they've been around for roughly ten thousand times longer than us. Let's just look at the numbers for a second:

- There are roughly seven billion (7,000,000,000) humans on Earth.
- Each human has roughly 30 trillion (30,000,000,000,000) cells.

- Earth has five million trillion trillion (5,000,000,000, 000,000,000,000,000,000,000) microbes, and that's just an estimate.

In fact, there may be as many as a trillion different species of microbes on Earth. So, it's not surprising that it is possible to find many, many more genetic adaptations within the microbial world than it is within the human world; it's simply a numbers game. However, we share a good deal of common genes with those microbes, and we even swap genes with them when we are in close contact (as happens in the microbiome in our guts), so it makes sense that we can learn and benefit from their adaptations.

Figure 2: Sharing Our DNA with Others

How much of our DNA (genes) do we share?	
With plants	18%
With yeast	26%
With the common fruit fly	44%
With the mouse	92%
With the chimpanzee	98%

The percentages shown in Figure 2 shouldn't be surprising, since we know that all living organisms share a single genetic code, so it stands to reason that common cellular functions between cells in different species would share common genes.

Given that microbes, plants, animals and humans share a huge number of basic biological processes and the DNA and proteins that control those processes, it is perfectly reasonable to ask, could any of the many adaptations that microbes have made over billions of years be useful to our own survival? Can we make use of that information or that altered or adaptive DNA for our own benefit? These questions are at the heart of many of the solutions that will be offered by synthetic biology.

Individual microbes (e.g., bacterial cells) have a great deal more in common with human cells than they have differences. It is therefore not unreasonable to think that a protein that was genetically adapted to protect a microbe against certain types of radiation (or any other form of stress) could be transferred to other organisms — including ourselves — to render them resistant to that radiation or other forms of stress. While this has never been attempted in humans, such manipulation of genes between cells, (e.g., different bacteria) has been performed, and so it is reasonable to think that such a transfer to human cells might be possible.

THE WONDERS OF MICROBES

Our appreciation and extensive use of microbes in biological research didn't come about by chance. Microbes are easy to work with and, given their commonality in biological processes with humans, they are ideal model systems for investigating novel pathways for everything from understanding basic biochemical and cellular processes (i.e., how we're put together and how we work) to developing new drugs. Microbiological research has been the basis for most of our understanding of gene function and the source of all our ideas in the area of "gene manipulation" that were developed over the past three or four decades. The rapid growth rates and short generation times of most microbial populations make them ideal research subjects and excellent models for the study of many common biological processes.

Microbes have often been considered Earth's most successful living organisms. That microbes adapt to changes — sometimes very extreme changes — in their environment has been known for many years. Whether the adaptations come about simply through random mutations within their genome or whether microbes are especially well-prepared to adapt to new pressures is less well-established and worthy of some comment.

In 1944, a graduate student at Columbia University named Evelyn Witkin made a serious error in her research that became a seminal experiment in biology. Overnight, she accidentally irradiated hundreds of millions of E. coli bacteria in a petri dish with a lethal dose of ultraviolet (UV) light, a form of radiation. When she returned to the lab the next morning, surprise! All the cells were dead — except for a tiny clone containing four bacterial cells that had apparently survived and continued to grow. Witkin had unknowingly developed a line of cells that were resistant to UV radiation, and she was smart enough to quickly ask an important question: Was the fact that some of the cells survived the lethal irradiation simply a coincidence, as a result of a random genetic mutation, or was the bacterium somehow "prepared" for the unexpected dose of radiation?

Witkin and her colleagues came up with the idea that the bacteria possess an "SOS" response, a DNA-repair process that they use when their genomes are damaged. The response involves the activation of dozens of genes and an increase in the rate of mutation. While these random mutations may sometimes be harmful, they also appear to enable adaptation to occur.

This repair process can be a good thing, as in this case with the development of resistance to UV irradiation. But in some situations, it's not, such as when bacteria develop resistance to drugs (antimicrobials), which has become a major medical problem. Consider, for example, the dreaded development of antibiotic resistance in hospitals — sometimes called AMR, for "antimicrobial resistance" — that is currently the bane of infectious-disease treatment. This gives rise to an ongoing question in biology that is of special interest to evolutionary biologists: Is this increase in mutation rate in bacteria merely a consequence of a DNA-repair mechanism that is prone to error, or is it an evolved adaptation that allows some microbes to survive serious insult or stress? There are obviously still some critical issues to be addressed.

Many people find it surprising to realize we have more so-called foreign microbes residing in our bodies than we do human cells. It's almost sort of insulting that we've been compelled to share our space, but we've developed a symbiotic relationship with those bacteria. We provide them with a place to live and prosper, and they provide us with protection so that we don't get infected by other harmful microbes. These microbes provide us with a sort of natural immune system.

The fact that microbes take up residence in our bodies is fascinating, and you might ask: How is this possible? How do these foreign cells avoid detection and rejection? The answer is that they have developed incredibly sophisticated strategies to avoid our immune system, which is programmed to get rid of them. Here are some of the strategies:

- Microbes practice biomimicry. They are genetically similar to the host individual (your body), which in turn is careful not to create antibodies that might also attack itself and therefore create an autoimmune condition.
- Microbes suppress antibody production using the strategy that the best defense is a good offense. In this case, they may inhibit the body's ability to respond to a particular microbe.
- Microbes hide within cells, where they can multiply and "prepare" to attack or simply prosper, all while avoiding detection.
- Microbes release antigens into the body's bloodstream to cause the organism to attack and destroy the antigens rather than the hiding microbes.

One of the takeaway lessons here is that, notwithstanding their tiny size, microbes are incredibly "smart" and have a huge repertoire that enables them to behave in myriad different ways. What's incredibly exciting is that we are increasingly coming to understand how these microbes achieve their "tricks," and we have even been able to identify which DNA and therefore which proteins are responsible. This has allowed us increasingly to understand basic biological principles and to develop new drugs to treat human disease when some of those biological processes go awry.

In addition to the microbes that are resistant to UV radiation or to antibiotics, there are microbial communities that have adapted to any number of environments. For example, living in:

- extreme heat close to or even within boiling lava flows;
- extreme cold, such as ancient glacial ice;
- extreme pressures up to 10 kilometers under the sea;
- extreme pH (extremely acidic or extremely basic), such as in high concentrations of arsenic acid;
- highly radioactive uranium; or
- any number of other extreme conditions.

Some scientists have taken to calling these microbes "extremophiles." What's really exciting is that by 2018, a good many (but obviously still a tiny percentage) of the genomes of these microbes had been sequenced and, in many cases, the gene or genes (protein or proteins) that allow for these adaptive behaviors have been identified and often understood. Now synthetic biologists are asking how they can use some of these characteristics to help us solve the critical problems that we increasingly face.

Author's note: If you're finding some of the terminology difficult, you might want to reference the glossary of terms and brief primer on genetics, which can be found following chapter 8.

HARNESSING THE POWER OF MICROBES

Let's look at an example to see how microbes might be used to help us deal with the "big three" challenges. In our example, a microbial community is exposed to some form of ionizing radiation.

The vast majority (more than 99.9999 percent) of the community dies from the radiation, but one microbe survives, quite by chance. It does so by mutating a gene that alters a protein that makes the microbe safe from the ionizing radiation. (It's probably a mutation involving the cell's ability to repair damaged DNA.) A new community of radiation-resistant microbes is thus established.

The question is whether we can identify that gene and introduce it into other organisms, including humans, plants or animals, in order to make them also resistant to ionizing radiation. The potential uses for that gene, or other gene products produced as a result of different environmental stresses, need be limited only by one's imagination. For example, as a general response to protecting mankind from radiation exposure caused by extreme thinning of the ozone layer, or perhaps as a way of enabling man or other animals to live on Mars, where the levels of ionizing radiation are thought to be high. Perhaps a better example might be to assess a microbe's ability to withstand cold and to incorporate its "antifreeze" proteins into corn or other grains to enable crops to withstand colder (or hotter, if the appropriate gene is discovered) temperatures, thus potentially increasing yields substantially.

Using the technologies associated with genetic engineering and synthetic biology to harness the wealth of genetic adaptations that microbes have undergone over billions of years creates the potential for solutions to some of man's most serious problems. The promise seems almost endless, and terrifically exciting. But you might be asking — and I would be surprised if you weren't — whether these technologies will allow us to go too far by "playing with" or changing life-forms in such substantial and possibly permanent ways. I would reply with two answers: The first would be that these are the sorts of things inherent in biology and the process of adaptation even without the help of human intervention. And second, without getting too philosophical, I would suggest that this is the sort of progress that humans have been encouraging on a continuous basis since the dawn of the first industrial revolution and even earlier.

A good example is the selective hybridization (a form of cross-breeding) of virtually all our food species (plants and animals) to better suit our needs, which include variables such as growth conditions, nutritional value, taste, shape, compatibility with transporting foods long distances from points of growth to points of consumption and any number of other properties. A distinct advantage that we have in 2019, which might have been absent even as recently as 2010, is the ability to make sure that all the genetic manipulations we make incorporate "safety switches" to ensure safety and, if necessary, containment.

And consider all the other things we do that appear none too natural. We move around our planet in airplanes; we explore other planets; we used to communicate with our voices whereas now we are more likely to use computer-generated bits and bytes; we use robots to do surgery . . . the list goes on and on. It is highly unlikely that humans will put a halt to any technology that might solve some of their most serious problems — for instance, with our health, food or environment.

EVOLUTION BY DESIGN

Another interesting question that arises is: Are we, with our new-found ability to manipulate specific genes without waiting for random mutations, looking at the demise of natural selection? That's a big question, and the simplest answer is probably yes; at least with respect to ourselves, plants and animals.

Juan Enriquez and Steve Gullans say it best in their book *Evolving Ourselves: How Unnatural Selection and Nonrandom Mutation Are Changing Life on Earth*:

> "If Darwin were alive today, he would likely revise a
> significant part of his great works, because the basic
> logic of evolution has shifted away from capital-n
> Natural toward two new core drivers: unnatural
> selection and nonrandom mutation."

Or, to put it another way, we are moving toward evolution by design rather than by natural selection. We'll be able to use genetic technologies, including synthetic biology, to select for specific traits in humans (for our health), plants and animals (for our food) and microbes (for our environment) and even to create completely novel life-forms to improve life on Earth. Without question, this is a very major event in the history of mankind. As you'll see below, it started off slowly, (e.g., using crossbreeding techniques to produce "better" foods), but today, it might increasingly represent an imperative if we are to solve some of the critical issues we face. Enriquez and Gullans state that:

> "We are transitioning from a hominid that is
> conscious of its environment into one that
> drastically shapes its own evolution."

If we examine the first several billion years of life's evolution — say, up until ten thousand years ago or so — the process of natural selection is pretty clear and is best explained by random mutations and survival of the fittest. Species survived and reproduced only if they were able to withstand the various environmental stresses to which they were exposed. So, as a matter of course, thousands, if not millions, of species (mostly microbes) would routinely disappear. Random mutations occurred continuously in all species but most would have no impact. Very occasionally, a random mutation might cause changes to a species that would give it a survival benefit, in which case it thrived even in the face of environmental pressures such as predators or extreme temperatures, such as the appearance and disappearance of ice ages. Without that random mutation and subsequent survivor benefit, the species would have dwindled or been driven to extinction. This is what we generally consider to be natural selection or evolution, according to Darwin.

There is evidence that as early as 9,000 to 11,000 years ago, humans practiced selective plant breeding and the effective "domestication" of agriculture. By using breeding to specifically manipulate the characteristics of particular species in order to alter or enhance their performance, humans were in fact already partaking in what Enriquez and Gullans refer to as unnatural selection via nonrandom mutations. Although this sort of activity has become more common in recent years with the development of ever-more sophisticated techniques of genetic engineering, the concept is not new. Later, farmers began to experiment with actual hybridization (combining different varieties of organisms to create a hybrid) to improve crop quality and quantity. This sort of activity was greatly enhanced by the work of the 19th-century Austrian monk Gregor Mendel, who is thought by many to be the father of the science of genetics.

GENE EDITING AND THE APPEARANCE OF CRISPR-CAS9

As soon as the original "genetic engineers" discovered how to clone genes in their laboratories and use molecular scissors (enzymes) to "cut and paste," the possibility of gene therapy became a goal for the treatment of a wide range of genetic diseases, such as muscular dystrophy and cystic fibrosis, and a host of other conditions in which "defective" genes are implicated. Unfortunately, for the most part it had remained a dream, as experiment after experiment either failed or fared poorly. We simply did not know how to come up with a workable method to edit out the bad gene and insert a good one, and not only to insert it but also to make sure that it was properly expressed. Scientists were mostly using quite artificial (non-biological, or at least non-natural) approaches and hoping they would work. For the most part they didn't, and in retrospect our approaches were pretty crude and really quite unrealistic.

Then came the discovery of CRISPR-Cas9, a naturally occurring gene-editing program present in certain bacteria that allows geneticists and medical researchers to specifically edit parts of a gene. The CRISPR-Cas9 system is a defense mechanism, rather like an immune system, that bacteria developed to protect against attacking viruses. Let's briefly look at how it works.

A virus invades a bacterium and incorporates part of its DNA (the viral DNA) into the bacteria's DNA. From the virus's point of view, that's a good thing; it's co-opting the bacteria's means of replication and establishing a home base from which to replicate. (Remember that viruses do not live on their own. They need a host home from which to thrive and replicate.) But it's not so good from the perspective of the bacterium, which in effect has been invaded, taken over and effectively been converted into a virus-generating machine. Luckily, its defense mechanism, the bacterium's CRISPR system, recognizes the intrusion and, using an enzyme that works like molecular scissors, it snips out the viral DNA and repairs the DNA strand to its original state. It even keeps a part of the viral invader's DNA, storing it in memory to defend against future attacks from the virus. Amazingly, this approach by the supposedly "lowly" bacterium is orders of magnitude more sophisticated than any forms of gene therapy that medical scientists have come up with in the past 40 years — further evidence of the sophistication and evolutionary development of simple microbes.

Scientists are now adapting this "natural technology" to attempt gene editing by cutting out sections of the DNA that may have faults or mistakes and substituting desired pieces of DNA in order to correct the fault. Medical researchers are editing out defective genes in various human diseases, including cancer. Food scientists are using the technique to improve the quality of foods by adding and deleting certain traits and environmental scientists are using

Figure 3: Gene Editing by CRISPR-Cas9

A DNA-editing technique called CRISPR-Cas9 works like a biological version of a word-processing program's "find and replace" function.

How It Works

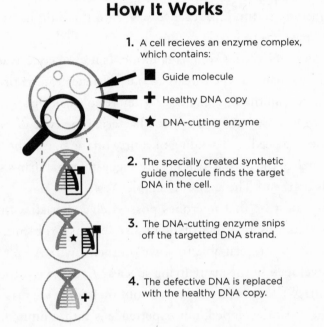

1. A cell recieves an enzyme complex, which contains:

   ■ Guide molecule

   ✚ Healthy DNA copy

   ★ DNA-cutting enzyme

2. The specially created synthetic guide molecule finds the target DNA in the cell.

3. The DNA-cutting enzyme snips off the targetted DNA strand.

4. The defective DNA is replaced with the healthy DNA copy.

gene-editing tools to better deal with issues in waste management for everything from food processing to oil spills to mine tailings. Perhaps equally exciting is the fact that the CRISPR-Cas9 system is probably just one of many gene-editing systems that bacteria have developed through time. Increasingly, these bacterial gene-editing processes are being discovered, examined and investigated with respect to their potential use in solving some of our critical problems, first in medicine and then subsequently in food and environmental issues.

What's critical about this methodology is that it's not about humans attempting to artificially correct biology. It's a biological tool developed over billions of years by microbes with the know-how to make those corrections in a biologically relevant fashion. Think about it for a moment. A simple microbe with the know-how to effect gene therapy in a manner that has failed many hundreds of brilliant medical scientists, dozens of well-funded biotechnology and pharmaceutical companies and government programs whose funding reached well into the billions of dollars over the past 20–30 years.

I know some of this sounds fantastic. Is it even conceivable that solutions to many of the threats to our existence can lie or at least partly lie within the tiny microbe? They're so small that you might be able to place as many as 150,000 bacteria (one type of microbe) on the tip of a needle, depending of course on the size of the bacteria and the size of the needle. Am I really suggesting that man's survival depends on them? The answer is simple: Yes.

How can it be that microbes possess characteristics that may lead to solutions to some of our most serious problems, which humans have not been able to solve on their own? Aren't we the most developed of the many living species? Aren't we at the top of the chain? Well, yes and no. It turns out that while we may appear to be the most developed, our experience is rather limited. We've been around for less than half a million years, whereas some of those microbes have been around for more than three billion years and, in many instances, they've experienced more and learned more over that time. Microbes have had ten thousand times more time to evolve and to adapt to conditions on Earth, and we'd be wise to learn from those adaptations and apply that knowledge to enable ourselves to adapt to some or even many of the pressures we find ourselves under.

Given the absolute revolution that we've been experiencing in genomic and information technologies, I believe that two essential ingredients are going to give us the ability to overcome the major challenges we face. Those ingredients are:

1. Our ability to recognize and define the way in which microbes and plants have evolved over billions of years and the "survival benefits" that they have achieved.
2. Our ability to adapt and shape our own evolution, and that of the plants, animals and microbes that support us, in the face of untold pressures.

No doubt this is scary stuff, but it's important to remember that these technologies can allow us to better understand the enormous and critical pressures that life on Earth faces and can provide us with solutions, unnatural though they may be, that may save mankind. My own feeling is that we have no choice. What's important is that we adopt these technologies in safe and appropriate ways.

Chapter Three

MAN-MADE

You might wonder how it is that we have arrived at the point where we have to learn from microbes and engineer novel biological entities in order to solve some of the most critical issues we've created. The simple answer is this: There has been a massive expansion of the human race which is impacting our planet, ranging from industrialization to urbanization to denuding Earth's forests. It has resulted in enormous environmental changes, and some have suggested that we have entered into a new, man-made epoch. In 2000, environmental chemist and Nobel Laureate Paul Crutzen gave it a name, the Anthropocene epoch, defined as:

> **"The current geological age, viewed as the period during which human activity has been the dominant influence on climate and the environment."**

MID-CENTURY INFLECTIONS

Some believe that the Anthropocene epoch began several thousands of years ago when man began to farm and live in specific communities (i.e., change the environment).

Figure 4: The Different Epochs

Development	Age	Epoch
Big Bang	13–14 billion years ago	Quark*
Earth is born	4.5 billion years ago	Hadean
Life begins	3.8 billion years ago	Archean
First plants	1.5 billion years ago	Huronian
Arthropods appear	590 million years ago	Paleozoic
Insects appear	400 million years ago	Paleozoic
Amphibians appear	300 million years ago	Pliocene
Dinosaurs appear	225 million years ago	Pleistocene
Humanoids appear	2 million Years ago	Holocene
Man changes Earth	8,000 years ago	Anthropocene

*The Quark epoch emerged 10–12 seconds after the big bang.

Others have suggested that the Anthropocene epoch really began around 1950, and when we examine the rapid rate of change and the growth curves that follow, it's clear that a number of interesting inflections, or tipping points, occurred sometime in the mid-20th century. (See Figure 5 below.)

Geologists studying the Anthropocene epoch have noted the appearance of radioactive elements across the planet following the development of the nuclear bomb, increased levels of pollution from plastics, soot from power plants, along with the disappearance of forests and the creation of "concrete jungles." Regardless of some dispute over the start date, many have accepted the notion of the Anthropocene epoch simply because humanity's impact on Earth is now so evident and profound. Much of that impact has put humanity at considerable risk.

In Thomas Friedman's *Thank You for Being Late: An Optimist's Guide to Thriving in the Age of Accelerations*, he lists a huge number of "Earth system" (physical) trends and socioeconomic (social) trends whose increase seemed to accelerate around the 1950 inflection point:

Figure 5: Physical and Social Trends Changing Over Time

Physical Trends	Social Trends
CO_2 in the atmosphere	Population
Ozone loss	Industrialization
Methane in the atmosphere	Urbanization
Nitrous oxide in the atmosphere	Fertilizer use
Domestication of land	Water use
Surface temperature	Primary energy use
Tropical forest loss	Real GDP
Ocean acidification	Large dams
Terrestrial biosphere degradation	Transportation and communication

As can be seen in Figure 6, a dramatic increase in the world's population started to become evident around 1950. The data demonstrate some of the dramatic changes that have occurred since then, and it's significant that a pretty sharp inflection in the changes appears following World War II, when rapid population growth and industrialization were taking place.

The labels pointing to Nobel laureates Fritz Haber (1918) and Norman Borlaug (1970) in Figure 6 refer to two critical events that occurred in the history of food production that effectively allowed population growth to continue. The first was Haber's discovery of a way (the Haber–Bosch reaction) to manufacture ammonia, a key ingredient in nitrogen fertilizer. Later, in the early '60s, Norman Borlaug — "the man who saved a billion lives" — developed winter wheat. Both advances helped to massively increase agricultural production worldwide and effectively allow for continued population growth. Those inventions along with the continued worldwide expansion of the industrial revolution that had started some hundred

Figure 6: World Population Growth, 1750–Present

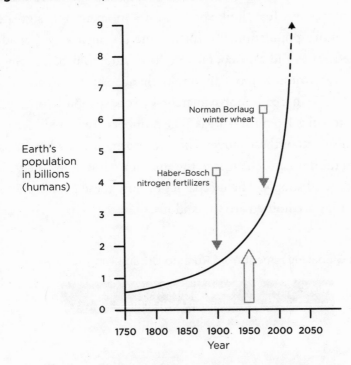

years earlier and the recovery from the devastating impact of World War II resulted in the huge increases in the world's population that continue today.

The issue is not just simple growth of the population but, more importantly, how and where those people are living. Following the war, there was a tremendous increase in wealth creation and a huge movement of populations from rural to urban settings. In retrospect, it should not be surprising that these two changes had an enormous impact on our food supply and the environment.

The urbanization that occurred in Europe and many parts of the Americas in the middle of the previous century is now occurring in Asia, especially in China and India. It is expected that by 2040 as many as (and probably more than) six hundred million people in both China and in India will have moved from subsistence living in rural communities to urban settings in massive and complex cities.

And the numbers moving in other areas of Asia and Africa are almost equally impressive. Just think about it for a moment: that's greater than the entire population of North America, including Canada, the United States and Mexico, moving from rural to urban settings. Another way to think about the rapid change is to recognize that in every week in 2018 approximately 550,000 people worldwide moved into cities from rural areas. The numbers are incredible, and the ramifications of these movements are enormous, nowhere more so than in the dramatic changes to the equations describing the security of the food supply. The mechanics of how these people are fed and how they are housed are stupendous.

Figure 7: Global Movement from Rural to Urban Living

Year	Percent of World's Inhabitants Living in Urban Settings
1900	8 (est.)
1950	13
1975	23
2000	43
2015	54
2030	67 (est.)

This straightforward urbanization would not be so serious if it did not coincide with a sharp increase in industrialization and major changes in energy use and their associated negative impact on our environment. Urbanization, along with a dramatic increase in gross domestic product (GDP), resulted in an increased demand for food for an upwardly mobile population. It is also associated with a concomitant loss of agricultural land, threatening food production and ensuing massive deforestation, resulting in a net increase of greenhouse gases in the atmosphere. This might just be one of the most serious ramifications of industrialization and urbanization.

Figure 8: Increase of GDP 1750–Present

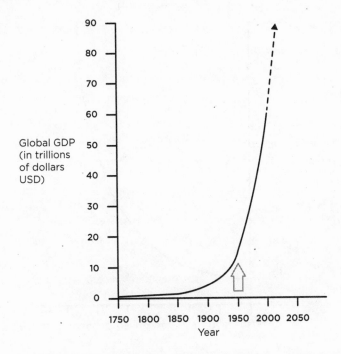

It is instructive — and again, in retrospect, really not surprising — that this population growth; this urbanization; this dramatic increase in industrialization, food production and energy use; the cutting down of our forests and a host of other factors have all coincided with a sharp increase in the production of greenhouse gases and, hence, global warming. For each of these, an inflection point is seen in and around 1950.

THE CARBON CYCLE

Let's focus for a moment on the carbon cycle, the huge acceleration in the burning of fossil fuels and the resulting "carbon problem." Prior to the start of industrialization (say, the early 1800s), the world's population numbered around or even a bit below one billion. During that

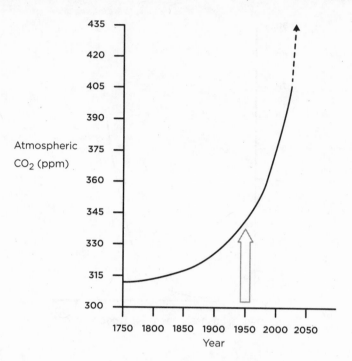

period, there was a sort of a harmony or equilibrium in the carbon cycle, with roughly equal amounts of carbon dioxide being removed from the atmosphere and used as the carbon source for all life on Earth (e.g., photosynthesis in plants) and then being returned to the atmosphere through the degradation of organic materials. Since the advent of the steam engine and the start of the second industrial revolution around 1870, the anthropogenic emmissions (there's that word again . . . Anthropocene) of carbon dioxide — largely through the burning of fossil fuels and the massive deforestation of the planet — have produced a major disequilibrium, with vast amounts of carbon dioxide increasingly being released into the atmosphere.

The term "greenhouse gas" refers to those gases in the atmosphere (water vapor, carbon dioxide, methane, nitrous oxide and ozone being the most important) that absorb and trap the sun's heat within the atmosphere, just as an actual greenhouse does. With an

Figure 10: Global Warming 1750–Present

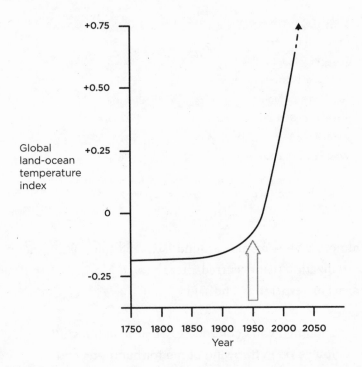

understanding of the atmospheric carbon dioxide measurements and of the various sources of increased carbon dioxide release, it simply cannot be refuted that global warming is largely the result of our dependence on burning fossil fuels, resulting in a sharp increase in carbon dioxide in our atmosphere. But it's also critical to understand that this is not just a carbon dioxide problem. The atmospheric levels of all sorts of chemicals exhibited similar inflection points around 1950 — not only carbon dioxide, but also methane, nitrous oxide and many others along with a thinning of the ozone layer. And change is not limited just to the atmosphere. Within that same time frame, our oceans have demonstrated significant acidification and warming, especially at upper levels. This, in turn, has had grave consequences for marine life and the health of the fishing industry around the world and, of course, represents yet an additional threat to the world's food supply.

Figure 11: Greenhouse Gas Emissions by Sector

Electric Power Stations	25.6%
Industrial Processes	15.9%
Transportation Fuels	13.2%
Land use, Biomass Burning	12.1%
Agricultural Production	11.6%
Fossil Fuel Retrieval, Distribution	10.5%
Residential, Commercial & Other	7.5%
Waste Disposal & Treatment	3.6%

Carbon dioxide (CO_2) is the major contributor, but methane (CH_4) and nitrouse oxide (N_2O) are also significant players.

It's almost as if, sometime around 1950 and the massive increase in industrialization that occurred after the war, Mother Earth threw up her arms in desperation and said:

> "Enough! I can't continue to absorb all the stuff you've been throwing at me for hundreds and thousands of years. Until recently, I was able to manage and take care of it all for you with little impact on your lives, but now it's simply too much and I can't cope. You are going to have to start doing things differently for yourselves."

In spite of all of the rational evidence, there are still those — often conspiracy theorists who also rail against genetically modified organisms (GMOs) and vaccinating children (the anti-vaxxers) — who question whether or not climate change is real and, if it is real, whether it's man-made. Despite what we've heard from President Trump and others, the evidence for climate change is overwhelming and so is the evidence that it is man-made. But even if it weren't man-made, we still have to deal with it if we are to survive. If there's a wall of fire heading toward your home, you know it's not a good idea to sit there arguing with your spouse about whether the fire was started by a smoker or by a lightning strike. That's just not a sensible use of your time or energy on such an occasion!

Figure 12: Acidification of the World's Oceans 1750–Present

Year	Ocean CO_2 (Mm/Kg)	Ocean pH*
1750	10	8.2
1800	10	8.2
1850	10	8.2
1900	10.5	8.2
1925	10.8	8.17
1950	11.1	8.15
1975	11.8	8.1
2000	12.1	8.05
2025	13.0	7.95
2050(?)	16.0	7.85
* The lower the pH, the higher the acidity		

This, to a certain extent, expresses the bad news: virtually uncontrolled population growth, huge differences around the world in terms of dire poverty and access to food, serious concerns that the world will suffer significant food shortages in the decades to come, untold increases in industrialization in most parts of the developed world and generally very little concern about the health of the environment.

There are some staunch believers who would have us undergo massive social and physical changes to our lifestyle to remedy these issues. They insist we must curb and even reverse population growth, stop burning any and all fossil fuels, halt mining, discontinue use of plastics and put an end to the cutting down of trees — in fact, they believe we must reverse deforestation and plant at least five hundred billion trees in each of the next five years. All of these are laudable suggestions, but unless countries around the globe undergo massive changes to their social and economic attitudes, many of these solutions are impractical at their core. And yet, I am not pessimistic. Quite the opposite, in fact.

There is good news: We appear to be starting a new era in which we are developing solutions to the problems caused by our actions over the past decades. This new era is sometimes referred to as the fourth industrial revolution and, as Steve Jobs predicted, the intersection of technology and biology is at its center. Please don't get me wrong — we're not getting a free pass. We will still have to make serious sacrifices and changes to the way in which we do business (population growth, burning fossil fuels, polluting the environment, etc.), but maybe we can do it with less pain than the measures cited above.

Chapter Four

A NEW ERA

As we enter this new era built on our expanded knowledge of biology and technology, the promise of synthetic biology is extraordinary, and it conjures up the words *visionaries, hopes* and *promises*. I quote from an obviously enthusiastic Craig Venter, one of the leaders of the Human Genome Project (HGP) and a pioneer of, if not the major leader of, synthetic biology:

> **"Over the next 20 years, synthetic genomics is going to be the standard of making anything. The chemical industry will depend on it. Hopefully, a large part of the energy industry will depend on it. We really need to find an alternative to taking carbon out of the ground, burning it and putting it into the atmosphere. This is the single biggest contribution I could make."**

So, what exactly is this new science or discipline of synthetic biology? Simply put, synthetic biologists design and engineer new biological entities to solve some of humanity's most pressing challenges in health, food and the environment.

Synthetic biology is not, of course, a science that stands alone; it uses technologies from many different disciplines, including biotechnology, genetic engineering, molecular biology, molecular engineering, systems biology, biophysics, electrical engineering, computer engineering, evolutionary biology and especially micro-biology and microbial genomics. It applies these disciplines to build artificial biological systems, in whole or in part, for research, engineering and virtually all life-science applications — and since we are in the midst of the Anthropocene epoch and its man-made challenges, it seems fitting that the potential solutions to those challenges should also be man-made.

GENETIC ENGINEERING

The development of ideas around genes and genetic manipulation started with the discovery of the structure of DNA by James Watson and Francis Crick in 1953, followed on by years of research on how genes work, how DNA and genes are synthesized and how they can be manipulated or changed. That is the basis of genetic engineering.

To diffuse any confusion in terminology, it's important to know that the term "genetic engineering" is synonymous with at least two other processes: genetic modification and genetic manipulation. Generally speaking, and quite arbitrarily, these processes would not include genetic breeding, whereby specific genes may be introduced into a species by crossbreeding (if it works or takes) rather than by the very specific manipulation of the genes.

Let me for a moment attempt an analogy. You own a car and it has four wheels, four doors, an engine that allows you to drive forward and backward, a braking system and a steering wheel allowing you to turn. But you have no direct access to any of those components; you can't modify them and you know nothing about how they work, only that they do, and you can drive the vehicle. Now along comes a "knowledge revolution," and you have a very deep understanding of how the car and of all of its various components work and, furthermore, you have the ability to engineer (manipulate or modify) any of those components. If the ride's too rough, modify the tires or suspension. If its top speed is too slow, up the speed (or keep it where it is for when your son takes the car). Put in a sunroof. Make the color of the car change according to the sunlight. Kill the starter switch if the car detects any alcohol associated with your son or any other driver. Have all of the instruments in the car respond to your vocal instructions (but not your son's). In fact, why not allow the car to drive itself? What a novel idea!

Thanks to the vast expansion in knowledge about biology, those are exactly the types of things that we can do today, even the kill switch. That's the power of genomics: to be able to get the blueprint for every living thing, how it's made and how it works. Now, I've exaggerated to make a point. We don't yet have the codes of *every* living thing; in fact, we don't even have the codes for 1/10th of 1 percent of them. But we do know how to get them, and when we do, we can understand how they function and can even reconstruct them when we want to and then adapt them when we have to. That is the power of DNA sequencing and DNA synthesis.

Take, for instance, a specific microbe that's been completely sequenced, so we know all of its DNA (and, therefore, exactly what proteins it expresses) and we have a pretty good understanding about the life of that microbe and which proteins allow it to do this, that or the other thing. Using the techniques of genetic engineering, synthetic biologists might want to manipulate the microbe's genes to suit their purposes. We might want to have the microbe improve the fermentation for alcoholic spirits, decrease the spoilage of vegetables on their way to the market, help sequester and remove lead and mercury from our polluted waterways or protect us from the scourge of antibiotic-resistant microbes

Or we might make a completely new organism by taking all of the genes (proteins) that we want to express, putting them all together with the minimum of other genes needed for the microbe's ability to grow and replicate. Bingo, we have a microbe that is especially generated through synthetic biology and genetic engineering to carry out whatever functions we are seeking: a microbe that captures carbon from the air, a microbe that delivers an antifreeze protein to allow plants to withstand freezing temperatures, a microbe that helps clean up oil spills — the possibilities are limitless.

On the one hand, the differences between the genetic engineering technologies developed over the past 50 or so years and the current thrust in synthetic biology may seem small. But, in fact, having the known sequences of many thousands (and that's just the start of it) of different genes saved on your computer and being able to synthesize brand-new DNA to engineer new biological entities to solve specific problems is revolutionary. For example, we can now think about creating a new food or a new plant or even an entirely new plant species by incorporating all of the desired genes (i.e., decrease the genes for fat and carbohydrate production while increasing the genes for protein composition in the final product, changing taste, texture, shelf life . . . there may be no limits). That is the essence of synthetic biology.

THE FOURTH INDUSTRIAL REVOLUTION

Clearly, synthetic biology is a very new, fashionable and transdisciplinary effort. But why now? Why has it exploded in such an extraordinary way? There are a number of critical reasons:

- We have very fast and very high-capacity computers that can help us to store (and retrieve) and synthesize DNA or genes responsible for a huge variety of different life processes.
- We have had a massive increase in our understanding of biology and how microbes and other cells function: how they do what they do; which genes are involved and how.

- We have the ability to sequence (read) the coded information stored in the DNA of all species.
- We now have the ability to synthesize (write) increasingly large chunks of DNA, up to and including entire genes.
- There has been a dramatic drop in the cost of doing all of the above, and the costs are only going lower.
- There has been a spectacular increase in our knowledge of biology and how organisms, especially microorganisms, have adapted and evolved (sometimes over a billion-plus years) to a whole range of physical stresses, many of which we are also exposed to.
- There is increasing recognition that we can learn from evolution to solve some of humanity's most serious challenges.
- And finally, there is the pressing reality that humanity faces some critical issues that do not yet have ready solutions.

In his brilliant book depicting the facts and implications of change, *Thank You for Being Late,* Thomas Friedman demonstrates how Moore's law (technically, the observation that the number of transistors in a dense integrated circuit doubles every two years) applies to the forces of technology and the impact they are having on so much of what we do as a people. This is certainly the case for synthetic biology, as the revolution that allows us to even think about this science has been built on the explosive development first of information technologies (circa 1975 to the present) and then of genomic technologies (circa 2000 to the present). It is critical to understand that synthetic biology didn't just happen; it came about as a result of the intersection of those monumental changes.

Computers are today at least a million times faster and have a hundred billion (yes, that's a hundred thousand million) times more storage capacity than they did less than a decade ago. This affords us the ability to rapidly amass, store and utilize the genetic codes of any number of different species (microbes, plants, insects and animals, including humans).

The capacity to store enormous amounts of information and access that information at incredible speeds is making synthetic biology possible. Remember that the human genome makes up approximately three billion bits of information and is not even the most complex of the genomes that have been discovered and sequenced to date.

Not only have the speeds and capacities of computers gone up sharply, but the costs to store data have decreased dramatically. In 1981, it cost $500,000 to store a gigabyte of data. By 2017, that cost had decreased to just $0.03 per gigabyte, a decrease of 16 million times. This is important because, in synthetic biology, we're going to have to be able to (1) collect, (2) store and (3) interrogate or search enormous bodies of data.

Figure 13: Data Storage Methods and Capacities 1975–2019

Year	Product	Gigabytes
1975	The floppy disc	0.00008 (80 bytes)
2005	Micro SD card	0.128 (128 megabytes)
2015	USB key	256 (256 gigabytes)
2016	Desktop hard drives	5,000 (5 terabytes)
2017	Google's cloud	150,000,000 (15 exabytes)
An increased capacity to store information of 2 trillion times (2,000,000,000,000) over 42 years.		

On the genome technologies side, the cost to sequence a human genome is 1/10,000,000 of what it was 18 years ago, and the time required to do the sequencing today is 1/100,000 of what it was as recently as 2000. That's a lot of zeroes, but I want to stress that the technological advances have been transformative, not just incremental, and this is what enables many of these technologies. Taken into context, while in the early 2000s it might have taken several years and many millions of dollars to discover a gene and sequence it, by 2019 the costs and times have decreased by six and seven orders of magnitude; that's up to 10 million times cheaper and faster. In 2000, not a single microbe had been sequenced, whereas in 2019, the number had already exceeded 150,000.

Figure 14: Changes in Costs to Sequence a Human Genome 2003-Present

Year	Cost	Time	No. of Genomes Sequenced
2003	$3,000,000,000	2 yrs.	1
2006	$20,000,000	6 mths.	1
2007	$2,000,000	1 mth.	1-10
2008	$200,000	2 wks.	20-50
2010	$10,000	5 days	200
2014	$1,000	1-2 days	500,000
2020 (?)	$100 (?)	<1hr. (?)	10,000,000 (?)

Has there been any other technology where the cost has gone down so dramatically and the uptake so rapidly?

In tandem with these developments, our knowledge of biology and its underlying genetics has also made massive strides since Watson and Crick's 1953 discovery of the structure of DNA. For example, we've identified genes and proteins that

- control cell growth rates
- cause cells to bind to mercury and lead
- allow cells to live at below-freezing temperatures
- enable cells to bind to and break down oil
- allow cells to absorb carbon dioxide
- identify mutated genes in cancer cells
- increase fat/oil production
- identify cancer cells
- allow cells to be resistant to ionizing radiation

And the list goes on, almost endlessly.

With all of this information plus our ability to read and store the sequences of the entire genome of thousands (even hundreds of thousands) of different organisms on our computers, the goalposts have changed dramatically. We can now synthesize DNA of increasing lengths — that is, synthetically produce genes that express any number of biological, cellular and chemical functions. This opens up a huge range of possibilities and the new field of science that is synthetic biology.

It should now be obvious that these advances in genetics and genomics accompanied by the massive changes in computer technologies have spawned a genomics revolution, which is only now starting to mature. We are now routinely seeing related headlines in magazine and newspapers. For instance:

"The next industrial revolution will come from nature . . . The combination of processes, principles and materials found in nature applied systematically to engineering is a new trend in manufacturing."

This is the headline of a July 2019 Fraunhofer press release, republished by the Science/Business Network. This is synthetic biology; make no mistake about it. The authors may stay away from using words like the feared and much-maligned phrase "genetic engineering," but that's exactly what it is. And I quote again from the same source: "new bio-intelligent manufacturing concepts will help to address the challenges we face."

SYNTHETIC BIOLOGY PIONEERS

In the coming pages I am going to call upon and champion the work of many different leaders in the field of synthetic biology. The field is new; we barely heard the term prior to 2014, but progress has been substantial. There are now many players in the game, both academics working in universities and synthetic biology engineers working in industry. Below, I introduce you to some of them.

SynBioBeta, the Synthetic Biology Innovation Network, is the industry hub for the young, vibrant field of synthetic biology. In the past five years, global annual investment in synthetic biology companies grew from $400 million in 2013 to more than $4 billion in 2019. In 2016, 12 companies raised more than $40 million each. Global publications in the field increased from barely a hundred in 2007 to more than 2,500 in 2018. Major tech founders are now investing in synthetic biology, including Bill Gates from Microsoft, Peter Thiel from PayPal, Eric Schmidt from Google and a host of other high-tech venture capitalists.

Five individuals have played especially important roles in the new fields of genome studies and synthetic biology:

Craig Venter is an American biotechnologist, biochemist, geneticist and businessman. In 1984, he arrived at the National Institutes of Health to examine how to identify messenger RNA (ribonucleic acid) in the brain. By the mid-90s, he was leading a private-sector initiative to sequence the human genome that was essentially completed in 2001. It was published in *Science* one day after the International Human Genome Sequencing Consortium published its first draft of the public-sector funded Human Genome Project in *Nature*. Venter is now the president of the J. Craig Venter Institute and an outspoken champion of synthetic biology or, using his terminology, "synthetic genomics." In 2005, he cofounded Synthetic Genomics, a company dedicated to using genetically modified microorganisms to produce clean fuels, among other products. In May 2010, he and a team of scientists at the Venter Institute became the first group to successfully create a fully artificial living cell.

George Church is a Harvard biologist as well as a geneticist, molecular engineer and chemist. He has worked in a whole host of different areas in genetics. In 2012, he published a brilliant book, *Regenesis: How Synthetic Biology Will Reinvent Nature and Ourselves*. One commentator has said that, while Charles Darwin explained how various species evolve, George Church "wants to turbocharge that process by putting new genes into organisms rather than waiting around for them to evolve those genes on their own." Church sees solutions to any number of the problems that plague humanity. In one approach, he used the revolutionary gene-editing technology CRISPR-Cas9 to alter 62 different pig genes in an effort to enable pig organs to be transferred into humans without being rejected. This would be a major breakthrough in addressing the critical issue of supply in the organ transplantation field.

Andrew Hessel is a young futurist and a catalyst of biological technologies, helping industry, academics and government authorities better understand the rapid changes happening in the life sciences. He is a distinguished researcher with San Francisco–based Autodesk Life Sciences, which is creating software tools for molecular and living systems. He is also the cofounder of Human Genomics, which is developing virus-based therapies for cancer. Along with George Church, Hessel is leading the second Human Genome Project, called HGP-Write, an effort to write (i.e., synthesize) and program the entire genome of humans and therefore any biological genome.

Juan Enriquez is a futurist, a visionary and an investor. As early as 2005, he was championing the revolutionary advances being made in biotechnology and the life sciences. His 2001 book, *As the Future Catches You: How Genomics & Other Forces Are Changing Your Life, Work, Health & Wealth*, is a tour de force that examines the future and the impacts of the biotechnology revolution. His investment firm, Excel Venture Management, was one of the first investors in Craig Venter's company, Synthetic Genomics. Enriquez's most recent book, co-authored with Steve Gullans and titled *Evolving Ourselves: How Unnatural Selection and Nonrandom Mutation Are Changing Life on Earth*, was highlighted in an earlier section (Evolution by Design) in Chapter 2.

James Collins is an American bioengineer, Termeer professor of Medical Engineering and Science at the Massachusetts Institute of Technology (MIT) and one of the founders and perhaps most prolific researchers in synthetic biology. Collins has been a consistent leader in the areas of biotechnology and biomedicine and in using the techniques of synthetic biology. He has achieved many significant breakthroughs, including paper-based diagnostics for Zika and Ebola that can be used inexpensively in the field in remote areas of Africa. In some of the most innovative approaches to infectious diseases, rare genetic abnormalities and inflammatory bowel disease, Collins has produced programmable synthetic cells to serve as "living diagnostics" and "living therapeutics" to both detect and treat these diseases that, to date, have resisted effective treatment.

EARLY SUCCESS

For some years, the discipline of synthetic biology motored along as a sort of exciting but somewhat exotic field of study. But then a critical event occurred in its evolution. On May 20, 2010, the English newspaper the *Guardian* published the following headline: "Craig Venter Creates Synthetic Life-Form." The scientific paper that spawned the article came out in print in *Science* several weeks later and was titled "Creation of a Bacterial Cell Controlled by a Chemically Synthesized Genome."

What Venter's group had essentially done was synthesize (chemically, without the use of any living biological material) a very long DNA molecule encompassing the minimum number of genes that would allow a simple bacterium to live and multiply. They then introduced that DNA into the shell of another cell (that is, a cell containing no DNA nor any of its own cell machinery). Within just a few doublings, the new cells were thriving, using only the information from the synthetic DNA. Venter's group, who obviously had a sense of humor, wrote four "watermarks" into the synthetic DNA: a code table for the entire alphabet, with punctuation marks; the names of the 46 scientists who had contributed to the project; three well-known quotations and a secret email address for the cell — all of that information chemically synthesized in the laboratory in the form of DNA code present on the computer. Just a decade ago, this creation of a new life-form would have been thought of as a pure episode of *Science Fiction Theatre*. Today, it's more like, "Wow, that's really amazing and it works."

Then, in 2016, Venter and his group reported on the creation of an artificial cell supported by a synthetic genome that contained the fewest genes (473) of any living organism. The aim of the project was to strip away all nonessential genes, leaving only the minimal set required to support and propagate life. In every sense, synthetic biology had become a reality and a novel synthetic life-form had been created.

With developments like these underway, it's exciting to consider some potential uses for synthetic biology apps, including:

- completely synthetic cells (à la Craig Venter above) used to synthesize vaccines cleanly and efficiently, perhaps even at the bedside;
- a microbe synthesized, either de novo or through genetic modification of an existing microbe, to deliver and fix nitrogen in a plant as a very specific and inexpensive nitrogen-based fertilizer;

- a microbe developed to convert corn or food waste into a completely biodegradable plastic polymer for any number of uses, such as water bottles, which could have massive implications for our energy and environmental sectors; and
- a gene-editing system such as CRISPR-Cas9 introduced into a cell to substitute a defective gene for a functioning one, for instance, for the treatment of muscular dystrophy or cystic fibrosis, or even a mental illness such as schizophrenia.

Even as the arena of synthetic biology is being developed, the question has been asked: Can we develop a cell-free system where we don't need the living (biological) cells but instead just use the engineered gene circuits in solution or even on strips of material, such as paper? Not only is this possible, but the cell-free technologies have significant advantages for environment issues. We can develop gene-based sensors for detecting toxins or other dangerous material in the environment, and they might even be used for remediation, cleaning up the various messes (e.g., toxic waste) that man has created. None of these examples are hypothetical. All of them are in an advanced phase of research; we'll discuss their future and that of others in the chapters to come.

SCIENCE FICTION OR SCIENCE FACT?

Before we go on, here is a scenario that could have been drawn from a *Star Trek* episode: The world is about to be exposed to a massive dose of ionizing radiation. Estimates are that within three to five years, 90 percent of the population will suffer from any number of severe, usually fatal outcomes. But stored in our computers is the entire genome sequence of a bacterium that is completely resistant to ionizing radiation. We have determined what gene, and therefore what protein, gives that microbe its radiation resistance. In fact, it's a DNA-repair enzyme that immediately corrects the mutation in DNA caused by the ionizing radiation. We take that gene and introduce it into the genome of a virus that has been rendered otherwise harmless, and then we direct the virus to the target — in this case, us. The virus infects all humans and has access to every cell in our body (think for a moment how you feel when infected by a flu virus). Once the protective gene has been delivered, the virus may be eradicated in the usual way we get rid of most viruses, or the virus can be programmed to self-destruct once it has carried out its mission. And now we are, just like the bacterium, totally resistant to the ionizing radiation.

I am well aware that I've been using words that smack of science fiction, but in today's new era, the potential for this is all real. These are the new tools of molecular genetics, genetic engineering and synthetic biology. Now let's take a closer look at how they are and will be used to remedy many of our biggest challenges.

Chapter Five

IN SERIOUS CONDITION

Let's begin with the good news. We've made major strides in medical research and huge advances in human health care. We've reduced the incidence and severity of cardiovascular disease through drugs, diet and exercise. We have vaccines to effectively control or even eliminate a variety of viruses such as polio and smallpox. We know how to combat many (but certainly not all) infectious diseases. We've developed tremendously sophisticated surgical interventions for musculoskeletal conditions, trauma and many other problems. It's amazing what surgeons can do to heal a broken body and what kidney, heart and other organ transplantation can do to save people suffering from organ failure. Figure 15 reflects these improvements by showing the decline in North American death rates per 100,000 population that occurred between 1960 and 2012. (Note that in some other parts of the world, especially in underdeveloped and recently developed countries, cancers and infectious diseases still go largely untreated.)

It wasn't that long ago (at least for us seniors) that we routinely heard that so-and-so had just dropped dead of a heart attack . . . "and he was so young." I'm not being politically incorrect here; it was youngish (45 to 55 years old) males who most often succumbed to heart attacks. Heart disease deaths have decreased by an average of more than 60 percent over the course of the past 50 years. We can attribute this to our improved understanding of the different forms

Figure 15: Deaths per 100,000 Population, Canada and the US, 1960 and 2012

	1960	2012	% Change
Heart disease	559	180	-68%
Stroke	178	39	-78%
Pneumonia	54	16	-70%
Smallpox		0	Eradicated
Stomach cancer	47	6	-70%
Pancreatic cancer	15	15	0
Lung cancer	20	31	+5%
Mental illness	3	14	+350%
Suicide/homicide	17.5	18.2	+5%

These numbers vary greatly between and within countries.
These represent a complilation of Canadian and U.S. data.

of cardiovascular disease and to the development of various treatments: statins for high cholesterol, beta-blockers and other drugs for high blood pressure, clot-busting drugs, surgical intervention for serious blockages or congenital abnormalities and, of course, disease prevention through dietary control and exercise.

In just 50 years, heart disease has gone from being a usually catastrophic event for relatively young men to a manageable chronic disease more often found in the elderly. Similarly, the rates of heart disease in women, in whom the onset of disease occurs later (most often post-menopause), has decreased as a result of understanding, prevention and medical intervention.

Acceptance of the germ theory of disease (from the days of Louis Pasteur), and an understanding of the importance of clean water and the development and widespread use of antibiotics brought about sharp decreases in deaths from infectious diseases. For instance, death rates from diphtheria, typhoid fever, measles, dysentery, whooping cough and scarlet fever have, at least in the Western world, been reduced to virtually zero. And, of course, some diseases have been eradicated either completely (e.g., smallpox) or "for the most part" (e.g., polio).

Between 1900 and 1980 the overall mortality from infectious diseases per 100,000 people in the United States decreased from 797 to just 36. The numbers in the rest of the developed world (Canada, Western Europe, etc.) are similar, although those in underdeveloped countries lag well behind.

The advances in surgery over the past century, though much less impactful (in total numbers) than the changes in death rates from infectious diseases, are almost miraculous. The earliest surgical interventions — neurosurgical, no less — occurred as early as seven thousand years ago. There is evidence that "surgeons" would drill into or remove a portion of the skull to relieve pressure on the brain following serious trauma or a stroke that involved uncontrolled bleeding beneath the skull. Surgical skills and tools obviously progressed over the years as knowledge of anatomy and physiology increased and the techniques and technologies available to the surgeon expanded. Currently surgeons regularly perform transplantation surgeries (whose success rate greatly improved with the development of antirejection drugs), laparoscopic surgery, microsurgery and even robotic surgery.

With all of these techniques, drugs, vaccinations, surgical interventions and other advances, you would think that the future of our health must look pretty rosy. Not only do we live longer and healthier lives, but there are also entire industries supporting that objective. It's true that we've solved or at least made major headway with these medical issues. However, without belittling the tremendous strides that we've made in combating a wide range of diseases, the fact is, we've been especially adept at picking what might be called the low-hanging fruit. Progress in many other medical conditions — for example, various cancers, neurological diseases, mental illnesses, certain infectious diseases and any number of genetic abnormalities — has been much less forthcoming, and several areas of human health remain in serious condition, with many of us still facing a huge number of medical issues for which there are not cures.

Given that the pharmaceutical and biotechnology industries had worldwide sales in excess of $1 trillion in 2016, and that's just for the drug part of the health care system, we have to question how costs will be met as we seek to solve the many outstanding medical issues. If you consider all the unmet needs in medicine in terms of the economic burden of pharmaceuticals alone — say, multiply that drug bill to $2 trillion or $3 trillion, which would represent 12–15 percent of total GDP — you can readily see how the system could quickly become unsustainable. Something will have to break in terms of cost structures. Fortunately, new technologies in synthetic biology have the potential to solve some of our most pressing health issues, as well as issues of cost and sustainability.

SYNTHETIC BIOLOGY AND HUMAN HEALTH

While health care is the "big three" challenge that appears to be least affected by the negative effects of the Anthropocene epoch, it stands to benefit significantly from man-made solutions that synthetic biologists are developing in the new epoch and new era. Why? Because synthetic biology offers an entirely new approach and perhaps even a more intelligent way to treat disease. In the older and of course still useful paradigm, a therapeutic target is established, and chemical or biological approaches are used to interfere with or modify the activity of the target (the site of the disease). Such approaches may be somewhat random; some use the expression "shotgun approach," as in the case of older cancer treatments, where it was hoped that the cancer cells may be more sensitive to a particular drug (toxic agent) than the surrounding normal tissue. Other treatments may incorporate various amounts of "biological intelligence" to deliver drugs specifically to the proper target. These approaches apply not only to cancer cells but also to infectious diseases (those caused by viruses and bacteria) and, to an extent, even to the diseases of the cardiovascular and nervous systems.

Synthetic biology is much more sophisticated. To quote one of the leaders in the field, Jim Collins of the Massachusetts Institute of Technology (MIT), "Synthetic biology brings together engineering and molecular biology to model, design and build synthetic gene circuits and other biomolecular components and uses them to rewire and reprogram organisms for a variety of purposes."

The range of possibilities now seems limitless, and activity seems to be increasing at an almost exponential rate in academia and in both small start-up as well as more established biotechnology companies. The following summary will be out-of-date before the ink is dry on this book, but it will give you a sense of the types of approaches that are being worked on.

Up until four or five years ago, it was possible to modify or edit genes and cells only in a test tube, but doing so in more complex systems, for example humans or a variety of animal models for human disease, proved much more difficult and even seemed impossible. Then along came CRISPR-Cas9, which showed us how bacteria perform these gene-editing tasks (addition or deletion), and we were off to the races. Even better, it's now clear that there are many different gene-editing systems in many different microbial systems that are yet to be discovered and repurposed.

The CRISPR-Cas9 systems in bacteria has captured the imagination of those investigating ways of fixing genetic mutations or abnormalities in a wide range of diseases, especially cancers and genetic diseases such as muscular dystrophy and cystic fibrosis. For example:

- Scientists in a laboratory in the Max Planck Institute in Germany are using a unique animal model, the tiny zebra fish, to demonstrate the possibility of using the CRISPR-Cas9 system to efficiently target and repair a specific genetic defect in the nervous system of the animal.
- A group at Temple University in the United States has successfully used the CRISPR-Cas9 system to remove

the AIDS-causing virus HIV from human immune cells. What's especially exciting is that the impact on the immune cells is permanent; not only do they become HIV free, but the cells are no longer susceptible to subsequent infection by the virus. In other words, the therapy imparted a sort of memory on the human immune cells so that they no longer allow themselves to be infected by HIV. Think about that for a moment: a simple little white blood cell with a memory.

- Many groups around the world are using the CRISPR-Cas9 system to try to solve the "mosquito problem" by finding a way to stop mosquitos from infecting people with the malaria parasite, which would be a major global achievement. By cutting out a piece of the gene responsible for the binding of the malaria parasite to the mosquito, the mosquito could no longer be infected by the virus. The intent would be to develop mosquitos that are resistant to malaria and therefore no longer able to spread the disease. Clinical trials are already in progress to determine whether this approach can be used to limit the spread of the Zika virus.

- A Swedish group is using the CRISPR-Cas9 system to edit DNA in healthy human embryos in an effort to decrease the incidence of miscarriages in patients undergoing in vitro fertilization (IVF) and to increase the rate of fertilization. This is strictly at a research stage.

Companies on the Cutting Edge

The CRISPR-Cas9 system has also captured the interest of many private-sector investors from early start-up companies as well as large multinational drug companies. Below are a few of them.

Intellia Therapeutics is a Boston-based biotechnology company that is using CRISPR-Cas9 technologies to develop specific gene-editing treatment for a variety of diseases, including inborn errors of metabolism and hepatitis B, and for the modification of hematopoietic (blood-based) stem cells prior to infusion.

Twist Bioscience is a San Francisco–based company that has become a major partner in the synthetic biology industry, providing tools that use their proprietary silicon-based processes for the production of long strands of synthetic DNA. They are also at the leading edge of using synthetic DNA to store digital data. Interestingly, DNA, which is based on a code that has four elements (A, C, G and T), has the potential to far outperform our current computers that use a code with only two elements (0 and 1). Not only that, but the information in DNA is stored in solution as opposed to on silicon wafers. It is estimated that a small test tube containing DNA could code for more information than the largest computer ever built. In 2013, a group of American scientists was able to store a JPEG photograph, the entire set of Shakespeare's sonnets and an audio tape of Martin Luther King Jr.'s famous "I Have a Dream" speech in little more than a teardrop of synthetic DNA.

Synlogic, in Cambridge, Massachusetts, was cofounded by synthetic biology pioneer Jim Collins. Starting with naturally occurring gut bacteria, Synlogic's scientists make use of the techniques of synthetic biology to reengineer or reprogram the bacteria to deliver medicine through the gut to disease sites. The bacteria may be engineered either to target infectious disease directly or to direct the immune system to do the specific work. The potential for spotting and stopping cancer cells, noxious viruses, bacteria, including those that may be antibiotic resistant is enormous.

Editas Medicine is another Cambridge, Massachusetts, gene-editing company. Its focus is on diseases with specific genetic mutations. It uses technologies based on CRISPR-Cas9 to treat genetic diseases such as muscular dystrophy, cystic fibrosis and sickle cell anemia.

Cell Design Labs is a California company that has the expertise to reengineer the molecular machinery of a human cell to give it a specific therapeutic function. As an example, its researchers are directing engineered white blood cells to more effectively target and destroy tumor cells, in the case of cancer, or particular microbes, in the case of infectious diseases.

Allevi (formerly BioBots) is a company located in Philadelphia that specializes in the 3-D printing of live cells. In the firm's own words, "BioBots is building tools to design and engineer life. Our goals are to cure disease, eliminate the organ waiting list, reverse climate change and live on other planets." Now, that's clear and concise and just a bit over the top, but these people are serious. They have an incredible toolbox to help accomplish many of the objectives described in this book.

AstraZeneca, a leading multinational pharmaceutical company based in Cambridge, England, has announced substantial investments in CRISPR-Cas9 technology to evaluate new drug targets in a number of different animal models of human disease.

Novartis, another multinational pharmaceutical, has partnered with Intellia Therapeutics and is using the genetic editing capabilities of the CRISPR-Cas9 system to alter ocular stem cells in an effort to use stem cell therapy for the treatment of particular eye disorders, such as macular degeneration.

Some of the major gurus of synthetic biology, including Jim Collins from MIT, have come together to form Senti Biosciences, a San Francisco–based company that is developing synthetic biology tools, including novel gene circuits that can build next-generation cells and gene therapies that can "adapt, sense and respond" to biological aberrations (read disease), as in mutated DNA or rogue cells that are or have the potential to be harmful.

Keith Pardee of the University of Toronto (who trained with Jim Collins at MIT) has begun to take the field of synthetic biology a step further by "hosting" synthetic gene networks in cell-free paper-based systems. This revolutionary technology has allowed Pardee to create a reliable test for the Zika virus — readily adapted for other viruses — that can be used at the bedside or in the field, without the necessity for sophisticated hospital- or laboratory-based equipment and interpretation. The RNA- or DNA-based sensor is incorporated into a piece of freeze-dried paper the size of a postage stamp. A very small sample of saliva, urine or blood is applied to the sensor, which yields a response in a short period of time. That is, the paper turns a certain color in response to the presence of a particular virus, not unlike the glucose test strips that many diabetics use to monitor glucose levels, but in this instance the sensor recognizes a bit of mutated DNA. The system is obviously very portable. In the case of the Zika virus, it enables reliable testing in the field at a relatively low cost — as little as one dollar per test. And with time and increased volumes, the cost should go down even more.

Note that we're way past the "idea" stage — these products are all in development. They have all passed the idea and discovery phases of research, have moved to the proof of principle stage and are well on their way to human clinical usage or at least trials. Many will no doubt fail, but there will also be spectacular successes.

And now for something that's nothing short of fantastic, a little eerie and maybe more than a bit hard to fathom. Scientists at the Columbia University Medical Center claim to have built the world's smallest "tape recorder" — inside a living bacterium. Mike McRae from ScienceAlert describes it: "Researchers have hacked the immune system of a bacterium into serving as the equivalent of a molecular tape recorder. By responding to chemical changes in the surroundings and then 'time-stamping' them in DNA, the technology paves the way for living monitoring devices that could be used in health screens or to analyze pollutants in ecosystems." This may not seem quite so fantastic if you recall the earlier discussion of the spectacular amount of data that can be captured in a single drop of DNA.

Think for a moment about how audio-visual equipment acquires and stores signals (on tape or a memory stick, for example) that can be played back to offer a historical accounting of events. Now, instead of monitoring a visual or auditory signal, what if you could monitor over time a molecular signal inside your body, inside a cell or maybe even inside your brain? The monitoring device is a bacterium (the equivalent of the camera or audio recorder), and the storage medium is DNA (the equivalent of the film or audiotape or, if you like, the digital equivalent). The Columbia scientists came up with the idea of using the gene-editing system CRISPR-Cas9 to continuously monitor changes in DNA and store those changes in a bacterium such as E. coli. They call the technology that monitors and records the changes TRACE, for "temporal recording in arrays [of DNA] by CRISPR expansion." The implications are phenomenal. The recorder might be able to monitor arthritis in the joints of your hand as a function of time or copper or mercury in your body as a function of diet and time, or it might monitor the fluctuation of specific disease markers in your digestive system. It might even be able to compensate for natural memory loss.

The potential for synthetic biology to create new "vehicles" to address various aspects of human disease really knows no boundaries. We can think of anything from single-purpose killer viruses that specifically engage DNA mutations in cancer cells to kill and destroy and then disappear once their mission is accomplished to microbots, microbial robots whose function it is to take up residence in the microbiome in your gut or airways and protect you against harmful bacterial, viral or microorganic invaders. While these approaches are no longer the fanciful ideas of science fiction, not all of them are quite ready for prime time. Nonetheless, let's take a closer look at how they and other synthetic biology solutions may help us to solve some of our most urgent medical issues such as infectious diseases, neurological and mental disorders and cancer.

INFECTIOUS DISEASES

Following the old adage "an ounce of prevention is worth a pound of cure," the treatment of infectious diseases with antibiotics and their prevention with vaccines have certainly been among the most important advances in medicine. Penicillin and dozens of other antibiotics, which are used to treat microbial (bacterial) infections, have had an extraordinary impact on mankind and our ability to combat infectious agents.

Another great advance in health care was the development of vaccines against infectious diseases such as smallpox and polio and a host of other less threatening diseases such as measles and mumps. Through enormously effective smallpox vaccination programs worldwide, the disease was officially declared eradicated in 1980. Prior to the development of the first smallpox vaccine — crude versions were being developed in England as early as the 17th century — smallpox was responsible for as many as five hundred million deaths.

The history of polio and the development of an effective vaccine, though not quite as dramatic as that of smallpox, also represents a huge win for disease prevention through vaccination. Polio is a viral disease that attacks the central nervous system, causing paralysis or death in a significant percentage of those affected. The first known polio epidemics were recorded early in the 20th century. In 1916, when a polio epidemic was declared in Brooklyn, New York, more than twenty-seven thousand cases were recorded in the United States, resulting in more than six thousand deaths. The first successful vaccines for this devastating disease were developed sequentially by Hilary Koprowski, Jonas Salk and Albert Sabin in the late 1940s to mid 1950s. The vaccines quickly took hold in the Western world, with developing countries following closely behind. In the United States, the number of polio cases decreased by over 90 percent in just over a year from fifty-eight thousand in the year before the Salk vaccine was developed to just fifty-six hundred cases in 1957. There have been no reported cases of polio in Western countries over the past decade. Such is the excitement and potential effectiveness of vaccination, but as the saying goes, "It takes a village," and the current anti-vaccine drives seen around the world are very worrisome. More about that in the final chapter.

The development of such achievements is relatively straightforward. First, the disease, its characteristics and its progression are investigated. What is the pathogen causing the disease? Usually it's a microorganism such as a bacterium, virus, parasite or fungus. In a patient who has the disease, what are the tools to combat the infection agent — antivirals, antibiotics, the patient's own immune system or, as is quite often the case, just letting the disease run its course? Then another important question is asked: How is the disease spread — by water, by air or through contact? And then: How can the disease and especially its spread be prevented? We wash our hands; we cover our mouths when we cough; we seek to eliminate the pathogens, often by vaccination.

The advances have been quite spectacular, but the challenges are still considerable. Even though we've been effective at eradicating some terrible diseases by vaccination, why can't we do it for other infectious diseases, whether viral or bacterial? Every year the influenza virus kills up to five hundred thousand people worldwide while annual flu shots (vaccines), good health and symptom management, including antiviral therapy, keep the disease from killing more. Yet, pandemics still occur: the Spanish influenza of 1918 killed 20-50 million people, the Hong Kong influenza virus killed a million people in 1968. Other viruses, such as HIV/AIDS and Ebola, are spread rapidly through contact with blood or other body fluids from an infected individual. Globally, the HIV/AIDS pandemic has, to date, killed more than 30 million since 1960, and the more recent Ebola virus epidemic of 2016 killed more than 11,300 people in West Africa. In poor areas of Africa, these diseases are almost invariably fatal, yet for a whole host of reasons, medical science has not been able to develop an effective vaccine. As this book was being finalized in the Spring of 2020, the COVID-19 virus struck the entire world. We were in lockdown for many months and the virus had already killed almost 400,000 people.

Vaccinations are meant not only to prevent sickness due to an infectious microbe but also, and perhaps more importantly, to limit the spread of that infectious agent within the surrounding population — the so-called issue of "herd immunity," whereby if enough of the population is inoculated (vaccinated), then all of the population is protected simply because the spread of the disease or virus is halted. When you are exposed to any foreign entity, especially a disease-creating microbe, your body mounts an immune "attack" on the bug, which means it produces antibodies to try to clear the microbes from your body. A "battle" ensues, and most often your body's antibody response wins out and you get better — or not even sick in the first place. In rare instances, your antibodies are not strong enough and the bug wins, causing anything from some discomfort (such as a cold), which might then run its course and disappear, all the way to death (as in Ebola).

The objective of vaccination (also referred to as immunization) is to prevent the microbe or disease from taking hold in the first case. Vaccines are produced by taking small parts of the microbe (antigens) or even the microbe itself in a "killed" or non-infectious form and injecting them into a healthy individual. In a sense, your body is being tricked into creating antibodies to mount an attack on or create a defense against a presumed disease-causing microbe. This antibody protection can last for years so that when you are exposed to the infectious agent, you have a defense ready and are spared the illness.

The list of diseases against which vaccines have been developed is impressive. It includes smallpox and polio, diphtheria, hepatitis A and B, influenza, measles, mumps, varicella (chickenpox), rubella (German measles), HPV (human papillomavirus, a cancer-causing virus) and a host of others. But compared to some of the sophisticated techniques of genetic engineering that we've discussed, the process of vaccine production is rather crude and frequently not successful. The first step in creating a vaccine is to generate one or more antigens that will lead to the immune response. This is done by growing the pathogen (the virus or bacteria) in large quantities in biological cells, either chicken eggs or specific lines of human cells. When enough of the antigen is produced, it is isolated and purified. (The antigen may be a simple molecule or the entire pathogen in an inactive form.)

The "magic" — or, rather, the uncertainty and, hence, the difficulty — of the vaccination process is that you're never sure whether the vaccine is going to produce enough of an antibody response in the recipient to stop an infectious agent from taking hold and causing significant disease. Literally tens of billions of dollars have been spent in an effort to create a vaccine against malaria, HIV/AIDS and Ebola, for example, with little to show for it. It almost seems as if the viruses have some sort of protection or strategy to avoid detection and destruction by developed vaccines. In fact, that probably is the case. They may be tiny microbes but they're smart. They may be avoiding effective vaccination either by continuously changing their surface antigens, against which the vaccine works, or by actually suppressing the immune system whose function it is to detect and destroy the pathogen. Which of these theories is true is not yet clear.

In the case of the influenza virus, the major limitations to traditional vaccine development are that it takes too long to identify the antigen against which an effective immune response will be mounted and too long (in terms of the flu season) for enough of the vaccine to be produced, particularly if the viral antigen changes every year. In 2009, for example, a particularly nasty influenza virus, H_1N_1, produced a pandemic. It took more than six months to create a vaccine, by which time the outbreak had largely subsided, and estimates are that between 150,000 and 575,000 people died worldwide. Similarly, as this book was being finalized, the entire world was eagerly awaiting the successful development of a vaccine against COVID-19.

Synthetic biology offers a cleaner and faster future for the production of vaccines. Success starts by knowing how to read and write the entire genome of the pathogen, whether it's a virus or a bacterium. Even if you don't have the code already in your computer, you can isolate and sequence the pathogen's entire genome for relatively few dollars in a matter of hours. (Remember that not too long ago, three to five years, this would have taken weeks if not months and at substantially higher costs.) In the next step, a chunk of DNA expressing the specific trait you want the vaccine to combat is transfected into mammalian cells, or even into synthetic cells whose sole purpose is to produce the vaccine. These cells then produce copious quantities of the vaccine as quickly as they can replicate, and they do it with very little need to separate and purify the result. We are talking about creating substantial amounts of an effective vaccine in days or even hours rather than weeks or months.

The process described above has now moved well past the research lab and into advanced clinical trials. Well-established pharmaceutical companies (such as Novartis) that have long been involved in the development and production of traditional vaccines are now working with synthetic biology companies (such as Craig Venter's Synthetic Genomics) to bring novel vaccines to the market quickly.

As we move forward, we may experience a bottleneck in convincing the regulatory agencies that synthetic biology processes produce safe and effective vaccines. However, I am not too concerned about this issue, because, as with many items under regulatory control, success breeds success. As these vaccines are shown to be effective and safe, approval rates will go up and the times to approval will go down. And on a related note, I am certain we will hear an outcry from the anti-vaccination crowd (anti-vaxxers), who already object to existing vaccines in spite of overwhelming evidence that their concerns are unfounded. I talk more about these people and the detrimental effects their false beliefs have on society in chapter 8.

Antimicrobial Resistance

Antimicrobial resistance (AMR) is becoming an increasingly serious problem in modern medicine as a result of today's widespread use of antibiotics. The term refers to the ability of microorganisms such as bacteria, viruses and some parasites to generate resistance to antimicrobial drugs. Given the ability of microbes to quickly adapt genetically to different environments, it is not a surprise that bacteria will adapt to resist antibacterial agents. It becomes serious, especially for hospitals, when an infection becomes resistant to *all* antibacterial drugs. The Centers for Disease Control in Atlanta reports that in the United States there are over two million antibiotic-resistant infections annually, leading to more than 23,000 direct deaths.

Because of the widespread use of antimicrobials to combat and/or reduce infection rates, not only in hospitals and households but also in agriculture (to reduce bacterial infections in many plants and animals), microbes have increased opportunities to adapt to the drugs and, for reasons that are not entirely clear, few new classes of antimicrobials have been developed over the past few decades. It would appear that the extensive repertoire of antibiotics that the pharmaceutical industry has produced is more frequently being combated by the ability of microbes to develop AMR. In response, agencies such as the World Health Organization are increasingly encouraging limited use of certain antibiotics, especially in nonhuman situations, to avoid the development of AMR in large microbial populations. If you think about it for a moment, every time you use an antibiotic for any purpose, you run the risk of encouraging a bacteria to develop resistance to the antibiotic and therefore stimulate the appearance of bacteria/microbes with AMR. Yes, it is frightening, and it's the reason we're continuously being warned against the overuse or inappropriate use of antibiotics.

Microbes' ability to adapt and become resistant is certainly not new; in his 1945 Nobel Prize lecture, Sir Alexander Fleming said,

"It is not difficult to make microbes resistant to penicillin." What is of such great concern now is that the overuse and misuse of antibacterial agents across the world are making the incidence of AMR more prevalent and more dangerous. Concern has been expressed that normally well-controlled bacterial infections could start to turn lethal more often as the microbes become resistant to any and all available antimicrobial agents.

Synthetic biology is likely to play a very specific role in combating AMR. Rather than relying on traditional antimicrobial agents, which microbes seem to be able work around or develop resistance to, synthetic biologists are reengineering bacteria and even producing synthetic cells (à la Venter) to create synthetic microbes that will be able to monitor their environment and destroy potentially dangerous infections. Examples might include:

- Using the very same tools that the microbe uses to develop its drug resistance in the first place. You can engineer bacteria that have been proven safe to "seek out and destroy" rogue bacteria that have developed resistance to all the common antibiotics. Your microbiome (your whole family of microbes, residing largely in your gut) could be enhanced to include these friendly bacteria that would go after disease-causing microbes.
- Generating genetic elements that could remove the genes from the bacteria that enable it to develop drug resistance.
- Attacking the very genes in bacteria that allow them to develop drug resistance. So, imagine for a moment that we can populate our mouths, lungs, skin and intestines (sites of major bacterial presence) with "defending bacteria" that determines which "invading bacteria" are able to develop AMR and then destroys those specific bacteria.
- Using synthetic genomics to create to generations of antibiotics to which bacteria cannot become resistant.

With these tactics, we could effectively put an end to the serious threat of AMR in humans, and there is no reason that we couldn't do the same for plants and animals. Remember also that AMR is not only developed by the overuse of antibiotics in humans, but the same problems can be seen in the use of antibiotics in plants and animals, given the fact that both the bacteria and the antibiotics may be involved in human disease as well. Note that when we talk about growing food without antibiotics, it's not that the antibiotics themselves are bad for you, but rather their overuse encourages the creation of micobes that are antibiotic resistant or AMRs.

It is critical to remember that we coexist — we need them as much they need us — with a broad microbial community that resides in various parts of our body (gut, lungs, skin, etc.), so any antimicrobial strategy has to be selective. A synthetic biology approach will allow us to target the specific mechanism that confers antimicrobial resistance to specific bacteria, or to other microorganisms that are unfriendly and require eradication.

DISEASES OF THE NERVOUS SYSTEM

For neurological diseases such as Parkinson's, multiple sclerosis, Alzheimer's, amyotrophic lateral sclerosis (ALS, or Lou Gehrig's disease) and for mental illnesses such as schizophrenia, depression, bipolar disease, obsessive-compulsive disorder, addiction and others, the prognosis for the future using current technologies is poor; actually, dismal is probably a more appropriate adjective.

Multiple Sclerosis

Let's briefly review multiple sclerosis (MS) as an example of many of the diseases of the nervous system. MS was described as early as 1838 as a debilitating disease of the nervous system, encompassing symptoms such as blurred vision, tingling and numbness in the extremities, fatigue, memory and concentration problems and, with time, difficulty walking. Even with its official designation as a disease of the nervous system in 1870, the symptoms were not always clear, and progression of the disease was extremely variable, ranging from occasional attacks and long periods of remission to acute and debilitating permanent attacks.

In 1960, MS was recognized as a demyelinating disease, whereby the sheath or covering of neurons in the spinal cord and elsewhere is attacked. By the early 1960s, it had become clear that MS could be linked to the immune system. It was thought that it might be an autoimmune disease in which the body's own immune system attacks the myelin sheaths that "insulate" the neurons. By the 1990s a series of drugs, all in the category of immune-based therapies, were being used with some success to treat a form of MS called relapsing-remitting MS, or RRMS. More drugs in this category have been developed, with moderate success, primarily in the RRMS form of the disease. But let's be clear: while exciting, these developments are relatively minor in terms of any sort of cure. The diagnosis of MS is still a devastating event. We still have no idea what causes the disease or, if it is an autoimmune disease, what the trigger is. Viral causes have been largely ruled out. Over the years, genetic causes have been implicated, but this approach has yet to bear any fruit. In spite of 50 years of intense study, we know little about the cause of this chronic, debilitating, somewhat unpredictable and often fatal disease.

I use MS as an example because it is reminiscent of many of the chronic diseases that are categorized as neurological, neuromuscular or diseases of the brain. Many basic and clinical approaches to understanding and treating diseases such as MS continue, but for the most part we are left with unclear and disappointing results, and patients, families and even researchers end up going through cycles of research, promise, hope, then despair.

The fact is that while there is always hope, the numbers are not good. If for the moment we lumped all of the neurological disorders together and asked for a status report, we might conclude:

- The amount of research dollars being invested both publicly and privately has been increasing substantially over the past several decades.
- Our ability to diagnose and understand the basis of the diseases has improved only very modestly over the past several decades.
- The incidence of the diseases for the most part has not changed substantially, with a few exceptions — most notably addiction, which has increased, as has the rate of suicides in those, especially youth, suffering from mental illness such as schizophrenia, bipolar disease and the results of addiction.
- There are no cures, only hope, and often despair, for those who are living with neurological disorders and their families.

Mental Illness

An equally or even more disheartening discourse might be offered for where we stand on the understanding and treatment of mental illness. These diseases, in addition to having devastating impacts on the affected individuals and their families, can also have enormous economic and social impacts. Based on data from 2010, it was estimated by the US National Institutes of Health that, worldwide, the direct and indirect cost of mental illness was US$2.5 trillion (now approaching $5 trillion). Unlike cancer and cardiovascular disease, where the larger fraction of costs is reflected in direct patient care, in the case of mental illness the indirect costs ($1.7 trillion) are twice the direct treatment costs ($0.8 trillion). These costs include the non-medical costs of the disease, such as disability, care-seeking, special housing, lost income and the like.

When dealing with conditions such as depression, anxiety disorders, bipolar disease and schizophrenia, as well as substance abuse or addiction, the root causes are largely unknown. A common review of mental illness treatment would invariably talk about chemical imbalances in the brain. Our understanding of those imbalances is poor at best, and over the years, only very small groups of drugs have been developed. These include antidepressants and anxiolytics (anti-panic or anti-anxiety drugs) for the treatment of acute anxiety; mood stabilizers, largely for mania; and antipsychotics for severe psychotic disorders such as schizophrenia, with lower doses being used to treat milder forms of anxiety. While all of the drugs have some scientific rationale for their use — for example, selective serotonin reuptake inhibitors (SSRIs) — it seems clear that much of the drug therapy used in the treatment of mental diseases is based on trial and error, as both the mechanisms of action of the drugs and the diseases themselves are often poorly defined. In addition to drug therapy, psychotherapy, behavioral modification, electroconvulsive therapy and deep brain stimulation are therapeutic approaches that offer some or occasional therapeutic benefit. What is disconcerting is that there have not been any "breakthrough" or fundamentally new drugs with different targets developed for the mental disease market over the past 45 years.

Interestingly, one of the most exciting recent developments in treatment for those who suffer from a variety of mental illnesses comes not from the biochemistry or physiology lab but largely from the information technology industry. That is, the use of "app technologies," largely in so-called talk therapy (for example, cognitive behavioral therapy, or CBT). Mental health apps can give the patient tremendous support with a variety of issues, ranging from compliance with medication protocols to 24/7 access to supportive or interventional therapy. This is interesting, even exciting but probably of limited use in the treatment of much of this very difficult family of diseases.

Gene Therapy for Neurological and Mental Disorders

For many years it has been postulated with some evidence that many of these disorders tended to "run in families." This has led to the notion that these diseases have an inherited component, in which case there would have to be a gene, or a number of genes involved. It is unlikely that we will find specific disease-causing genes but rather genes or series of genes that "predispose" an individual to mental illness. Nevertheless, ever since the discovery of DNA and the fact that certain inherited diseases are associated with genetic defects, or "errors" in the DNA, the notion of gene therapy has been top of mind for researchers. You may be thinking that if we have a "bad" gene and know what the gene is, and if we can make or derive a "good" or correct copy of the gene, we can simply put the gene in a pill or inject it to cure the disease. Unfortunately, it's not quite that simple. As with all therapeutic approaches, it's a matter of how, when and where the gene is to be delivered.

Assuming that we have the correct gene and that the disease involves a single gene defect, we have to consider: how is it to be delivered and how does it take the place of the defective gene? The challenges are many:

- In what form is the gene or DNA to be delivered?
- How does the gene and/or its product avoid being identified as a foreign substance by the immune system and, thus, broken down and cleared from the body by the liver or kidneys?
- If the gene has to get to a specific target (for example, muscles, as in the case of muscular dystrophy), how

does it find that tissue and get into the specific cells that express the genetic defect?

- Once it gets into the target cell, will the gene's expression be regulated in the same way that it is in normal cells? This is important, because if the gene is not regulated properly, then it might exacerbate the condition.
- Is the defective gene doing something that we don't want done, and if so, how are we going to stop that?

The list of questions goes on. The gene for muscular dystrophy was discovered in 1986, and since then, hopes for a cure using gene therapy have been high. But progress has been very disappointing; the barriers to getting the gene to affected muscle cells have so far been insurmountable.

Recent attempts to use gene therapy in cells that involve blood and the immune system have shown some promise. In these instances, blood cells — usually white cells that are immunologically active — are removed from circulation and genetically modified *ex vivo* (outside the body) to make them more effective at carrying out particular missions. This approach is gaining considerable traction, especially with the use of "natural gene therapy" represented by the CRISPR-Cas9 system that we described earlier.

However, dealing with diseases of the brain and nervous system is throwing additional challenges at us. Schizophrenia and muscular dystrophy are two devastating diseases of the brain and nervous system. They are very different in all respects but, just maybe, their resolution, their curse, might have a common thread.

Muscular Dystrophy

Unlike the case for virtually all other diseases of the nervous system, including mental illness, the molecular mechanisms associated with muscular dystrophy, especially Duchenne muscular dystrophy (DMD), are very well-known. The disease is a result of a very specific genetic mutation. The dystrophin gene and its particular locus on the short arm of chromosome X were discovered by Ron Worton in Toronto in 1985. But even though the specific mutation has been known for more than 30 years, there is no cure for DMD. While treatments have improved the quality and duration of life for some, the overall prognosis is still bleak, not much better than it was 50 years ago.

Knowing the specific gene associated with DMD, knowing in which tissue the defect is most usually expressed (skeletal muscle) and being able to clone the gene in quantity makes muscular dystrophy a favorite target for gene therapy. Indeed, gene therapy has proven effective in treating a mouse model of muscular dystrophy that suffers from the same dystrophin defect. But in spite of dozens of different approaches to gene therapy in human patients with muscular dystrophy, all of them have failed. Regardless of whether the gene was delivered through some viral vector (i.e., attached to a virus that is acting as a sort of "Trojan Horse") or through different types of cells, including stem cells, we have seen only failure. It is likely that the barrier to getting a sufficient number of copies of the correct gene to the muscle where the defect is expressed is simply too difficult to overcome. Like many failures in medical therapeutics, it may well be an issue of delivery, of getting the right treatment to the right place at the right time.

All this began to change in 2015, when a group of scientists on both US coasts discovered the CRISPR-Cas9 system and its potential for use in gene therapy began to be explored. The muscular dystrophy–related approach involves designing and building a synthetic virus that will deliver a corrected version of the defective dystrophin gene to affected muscle, along with a CRISPR-Cas9 system that will do the molecular surgery, cutting out the defective gene and replacing it with the correct one. By early 2019, clinical trials of this sophisticated biological treatment approach had begun.

Schizophrenia

Schizophrenia is a serious mental disorder characterized by antisocial and often psychotic behavior. The affected individual may not know or understand what is real, may hear voices and suffer from paranoia, and there's often comorbidity with anxiety, depression and substance abuse. Here are some words from a person suffering from the disease:

> **"My greatest fear is this brain of mine . . .**
> **The worst thing imaginable is to be terrified of**
> **one's own mind, the very matter that controls all**
> **that we are and all that we do."**

The fundamental basis of schizophrenia was until very recently largely unknown, and the treatment has been mostly with drugs originally developed in the 1950s and '60s as typical and atypical antipsychotics, with many similar or slightly modified drugs still in general use. It is exceptional, though maybe not so surprising, given the complexity of the disease and the lack of detailed knowledge of its causes, that very few new therapeutic approaches have been developed after so many years.

Schizophrenia — and the same likely holds true for other mental illnesses — is likely not a genetic disease in the traditional sense; that is, there is no single gene that causes the disease, nor are its inheritance properties clear. However, there has for many years been the notion of partial inheritability; that is, the disease appears to "run in families," but in very unpredictable ways. Now, following decades of investigation, there is increasing evidence from large population studies that schizophrenia is associated with particular genetic mutations that occur in many (even in as many as hundreds) of the genes associated with synaptic function (communications between nerve cells), as well as in areas of the major histocompatibility complex, a part of the immune system. While very little of the physiology has yet been worked out and there are now many identified genes involved, there is much excitement about the fact that we are beginning to understand, or at least to discover, some of the pathology and the basis of the disease and maybe even some new targets for therapy or intervention.

If schizophrenia is indeed associated with genetic mutations that alter the ability of nerve cells to transmit impulses or information between each other, synthetic biologists will attempt to create small viruses directed at the nerve cells or the synapses (the gaps between them). The synthetic virus's genome will be small and have a very limited repertoire. Specifically, it will contain a targeting element that will allow the virus to target cells where the genetic mutation is housed, a CRISPR-Cas9–like system that will cut out the mutated part of gene and replace it with normal DNA and possibly a "suicide gene" that will cause the virus to self-destruct once it's finished its mission. That sounds like a line from the film *Mission Impossible*, except in this case all of the elements are in the realm of the possible. That is the case for schizophrenia, but it holds equally well for other serious mental illnesses such as bipolar disease, severe depression, addiction and others. Certainly, these are very exciting times, but it is still very early.

CANCER

The status of cancer treatments is hugely variable and ranges from offering enormous hope and very positive initial results to hopeless despair. Many cancers are both complex and variable, with regard to both the form of the disease itself and the variability of the treatments used. Those treatments include:

- *Surgical removal:* This form of therapy works great when the tumor is "encapsulated" or solid and no cells remain as a result of being missed or because of metastasis (the spread of tumor cells beyond the primary site). In addition, surgery is sometimes used to reduce the tumor load prior to the use of some of the therapies that follow.

- *Radiation treatment:* This form of therapy is used to kill tumor cells and often to shrink tumor masses. It is based on the notion that radiation can kill the cancer while sparing untoward damage to the surrounding normal tissue.

- *Chemotherapy:* In a manner not dissimilar to radiation therapy, this form of treatment presupposes that the prescribed drug is more lethal to the cancer cells (if rapidly growing/dividing) than to similar or adjoining non-cancerous cells. The dilemma faced in both radiation therapy and chemotherapy is that there are many rapidly growing and radiation-sensitive cells in the body that are not cancerous.

- *Immunotherapy:* It is thought that in many cases where a tumor exists, your body may recognize it as foreign and attempt to mount an immune response against the offending cells, potentially disabling most if not all of those cells. However, too often the immune system is not strong enough to thwart the growth of a tumor and, in the ensuing battle, the cancer wins. Bolstering the immune system with drugs is an approach that has borne some fruit and is an important target for the use of synthetic biology.

- *Hormone therapy:* This therapy uses hormones to stop or slow the growth of tumors whose growth is hormone dependent, for example, in the case of breast and prostate cancer.

- *Targeted therapy:* This fairly modern form of chemotherapy identifies the differences in tumor cells that make them cancerous (allowing them to grow, divide and spread) and uses a specific drug to target cells that demonstrate those "cancerous" characteristics. This often, but not always, entails the use of specific antibodies targeted against a particular part of the tumor. This is the Holy Grail of cancer treatment: to create a magic bullet that does nothing but identify, target and kill specific cancers.
- *Stem cell transplant:* This is not so much a form of cancer therapy as it is replacing blood-forming stem cells (e.g., white blood cells) that might have been killed, either purposefully or not, during high-dose radiation or chemotherapy treatments. In a more recent twist, however, the stem cells might also be programmed to seek out and destroy the tumor.
- *Precision medicine:* The new gold standard for cancer treatment, which is still only in its infancy, is to select specific treatments that work efficiently against particular cancer cells. This approach frequently uses genomics to understand the disease and the difference between the cancer cells and normal cells, and then capitalizes on those differences to design the treatment.

These last three approaches, targeted therapy, stem cell transplant, and precision medicine — all of which are forms of immunotherapy — represent ways in which the techniques of synthetic biology can produce therapies, usually cell- or virus-based, that can be both very effective and very selective.

In 1992, Dr. Steven Rosenberg, a cancer surgeon at the National Institutes of Health in Bethesda, Maryland, performed an experiment that pioneered the entire area of immunotherapy and foreshadowed the potential of synthetic biology. In removing a solid cancerous tumor from a patient, he observed that the tumor was loaded with white blood cells, or lymphocytes, and he hypothesized that this patient's own cells had identified the tumor as something foreign, something that had to be destroyed. Unfortunately, while the lymphocytes were able to search out the cancer, they were not strong enough to kill it. Rosenberg then asked what would happen if he could extract the lymphocytes from the tumor and genetically modify or simply stimulate them in culture to make them more effective "killer cells" and then re-inject them into the same patient, where hopefully they would carry out their "search and destroy" mission more effectively. The experiment worked and the patient's cancer, in this case an advanced melanoma, went into remission.

The success rates with this approach were good but not outstanding. However, Rosenberg's work represented the first effective experiment in immunotherapy using genetic manipulation to produce more effective tumor-killing white blood cells. That was in 1992. Today we know so much more about the molecular (genetic) basis of identifying cancer cells and genetically modifying white blood cells to allow them to destroy the tumor cells more effectively. More recent clinical trials by Rosenberg and a host of other clinician scientists have ramped up both the search-and-identify and the kill-and-destroy potential of specific lymphocyte populations against a number of different cancers with terrific results. A recent report from the National Institutes of Health describes an immunotherapy approach targeting specific gene mutations found only in certain breast cancers that produced very high rates of long-term remission; however, the cancer community is often reticent, as it should be, to use the term "cure."

With all of this armamentarium available, one might expect that the battle to conquer cancer would be mostly behind us at this point. Not quite. The fact is, we have not been able to cure many of the difficult cancers. Sometimes we are able to achieve real remission — that is, no detectable cancer for a period of time — and sometimes the patient lives with the cancer as a chronic disease. But still, too often the diagnosis represents a death sentence, with just a modicum of hope.

The data with respect to cancer death rates and change over time are complex. They differ widely according to country, age and sex. Nevertheless, there are some trends that seem clear:

- The incidence of some cancers has decreased substantially (prostate, stomach).
- Death rates from some cancers have decreased in a significant but still modest fashion (breast, colon).
- Some cancers continue to have terrible prognoses, with little substantial progress in treatment approaches (lung, pancreas).

Novel Approaches to Conquering Cancer

In his State of the Union address in 1971, Richard Nixon declared war on cancer, stating:

> **"The time has come in America when the same type of concentrated effort that split the atom and took man to the Moon should be turned toward conquering this dread disease."**

There's no doubt that there has been a concentrated effort. Globally, we are currently spending about $120 billion per year on cancer drugs alone, and this number has been increasing at a rate of around 10 percent per year. In addition, in 2012 the US government spent in excess of $5 billion on cancer research and the biotech/pharma industry reported an additional $49.5 billion in oncology research. Together with other sources of funds, such as philanthropy, non-government agencies and other countries, the number must now exceed $100 billion per year — a situation that many consider unsustainable.

Regrettably, even with all these resources and the numerous therapeutic approaches available, many cancers remain lethal. Fortunately, and probably not surprisingly, scientists working in synthetic biology are looking at novel cancer treatments more than any other area. The major thrust in cancer-related synthetic biology is the engineering of a novel entity that is specifically designed to both detect the cancer cells and kill them. This may be an engineered synthetic virus, a totally synthetic cell (à la Craig Venter), a modified cell of our own immune system or even a microbe, all of which would have a specific search-and-destroy mission. The purpose of the new entity might be to eradicate the entire cancer cell, or it might be to render cells non-cancerous by specifically going after the genetic mutations that are causing the cells to proliferate. As is the major theme in this book, the approaches are largely "biological," often taking a lesson from how microbes do things.

A group from the University of California, San Francisco, led by Wendell Lim, has cloned a new form of a protein. It can be looked at as a switch that stretches from the outside of a cell — using a part of the molecule that can function as a sensor or cancer *detector* — to the inside of the cell — using a part of the molecule that acts as an *effector* or, to put it more dramatically, a cancer killer. So, "therapeutic cells" may be introduced into the body in response to detection of a particular cancer, or you may have "premeditated" cells circulating routinely, both with a search-and-destroy mission. This technology, of course, need not be limited to cancer. It could have the ability to recognize any molecule on the surface of any cell and to turn on genes to affect the appropriate response.

Andrew Hessel, one of the leading voices for synthetic biology, has described perhaps the ultimate in personalized targeted cancer therapy, an approach that relies heavily on genomics and synthetic biology. The basis for this approach lies in:

- fast, inexpensive sequencing of a tumor cell's genome and rapid detection of the cancer-causing genetic mutations (remember that in all cases, it's these genetic mutations that are causing the cancer);
- our understanding of a group of viruses called onco-lytic viruses, which have the capacity to selectively and preferentially infect and kill tumor cells; and
- our ability, following Craig Venter's work, to synthe-size an entire viral genome.

The methodology might be as follows. First, a small amount of the tumor is obtained. It could come from a solid tumor, some of which may have been excised, or a tumor in soft tissue, or it might be blood- or fluid-borne — the key is that some of the cancer cells be available. The tumor cells are sequenced, and the genetic mutation associated with the cancer is determined. (Today the sequencing would probably have to be done in a central laboratory, but with the development of new, inexpensive miniature sequencers, this whole procedure might even be done at the patient's bedside or in a physician's office.) The mutation is then read, and a viral genome is synthesized using newly developed DNA-writing techniques with the express function of recognizing, attacking and killing the tumor cells that possess that specific mutation. The new virus is quickly grown in cultured cells and then administered to the patient.

This is truly personalized medicine. The therapeutic agent (in this case, a virus) would be uniquely made with the intent of killing all the specific tumor cells in that particular patient. If necessary, the virus could even have a "kill switch" that would inactivate it after it has killed the tumor cells or when it is taken up by cells without the particular mutation it was created to fight. One of the keys to the use of viruses in the seek-and-destroy mission is their ability to reach virtually any cells, tissues or organs within the human body.

While this scenario may read a bit like science fiction, again, it has a strong basis in newly developed technologies that will make it possible, and not necessarily that far in the future. With the cost of sequencing and the times to sequence dropping so quickly, the option of making an inexpensive DNA sequence to function at the bedside is more than plausible. Companies such as Gen9 (recently acquired by Ginkgo Bioworks in Boston) and the Bio/Nano group at Autodesk in San Francisco are at the forefront of these approaches. Andrew Hessel, assessing the early work, comments:

> **"This work demonstrates that personalized made-on-demand therapies are within reach, and our efforts combating cancer in animal models could lead the way in next-generation care."**

And keep in mind, all these approaches need not be limited to cancer therapies. They may also be used to improve the targeting of antibiotics for use in infectious diseases or in the realm of agriculture, to protect food crops from unwanted predators.

Overall, there is still a lot more biology to be understood; after all, the human body is incredibly complex. But with the rapid pace of recent progress, it will not be long before we are routinely implementing the tools of synthetic biology to help cure infectious diseases, neurological and mental illnesses and cancer. It is clear that the fundamentals for new therapeutic methods are now available, and the promise of these advances I've just outlined is terrifically exciting.

Chapter Six

HUNGER

While issues of human health are difficult, issues of global food security — a sufficient supply of safe and nutritious food to feed the world — are even more complex. The complexity and the diversity of issues with regard to the food supply are perhaps best exemplified by these facts. In 2017:

- 150 million children were considered severely underweight or starving.
- Four million children died of starvation; that's one child every six seconds.
- 820 million adults suffered from life-threatening malnourishment.
- 12 million adults died of starvation; that's one person approximately every second.
- 110 million children were considered obese.

Just looking at these numbers, and especially considering the number of obese children, you might jump to the conclusion that the problem is surely not food production or the amount of food available globally. You might conclude that the issue is one of poverty and the availability of food, for both the starving and the obese. You might be right to say that we are dealing with an issue of access to food, not global food production per se. But you might also be wrong, and we can hardly afford to be wrong when assessing the world's ability to feed its people, whether poor or rich, obese or starving. There are those who believe strongly that there are serious limitations to how much food can actually be grown on Earth with traditional and even advanced technologies. Let's dive a little deeper before examining some solutions.

To start with, there are a number of different definitions of hunger. I've chosen a rather simple one:

"A condition whereby a person for a sustained period of time is unable to eat sufficient amounts of food to sustain basic nutritional needs, which in time will lead to death."

A plethora of organizations collect and distribute data on world hunger. Depending on their cause or bias, the glass — or rather, the plate — is either half empty or half full. Let's start with some good news. In 1996, the World Food Summit set out a specific objective: starting from a baseline in 1990–92 and ending in 2015, reduce by 50 percent the number of people in the world suffering from severe malnutrition. The results have been somewhat gratifying. In 1990, out of a global population of almost six billion, more than one billion people (representing 17.8 percent of the world's population) were considered chronically malnourished. By 2014, this absolute number had remained almost constant at 950 million malnourished, but with a population now slightly in excess of seven billion, the percentage had dropped to "just" 13.2 percent, a reduction of a bit more than 25 percent.

As you might expect, the distribution of this good news is not uniform around the world. Latin America has been the most successful in decreasing hunger over the period, dropping from 14.7 percent to 5.5 percent. In most parts of India and China, there was little change over that period, with levels of malnourishment much higher in India than in China. The picture in Africa is certainly the most depressing, with overall modest decreases starting from much-elevated levels (27.6 to 24 percent). The distribution of the changes in Africa are far from uniform, as we routinely see stories about starvation and death from starvation on the nightly newscasts from such countries as Nigeria, Yemen and Sudan. In parts of sub-Saharan Africa, there is evidence that the situation is growing grimmer year by year since 2014. There are two related cycles leading to these dire circumstances. On the human level there is poverty, population growth in the absence of any form of birth control, disease, malnutrition and death, and on the physical level there are adverse climate conditions, crop failures, drought, increasingly poor soil, deforestation and war.

While current statistics are concerning enough, some experts are forecasting a critical worldwide food shortage by the year 2050, which would clearly exacerbate existing problems. For example, in his book *The Coming Famine*, award-winning journalist and science writer Julian Cribb "lays out a very vivid picture of an impending planetary crises — a global food shortage that threatens to hit by mid-century — that would dwarf any in our previous experience," according to the publisher.

In any rational discussion of food security — also defined as "the state of having reliable access to a sufficient quantity of affordable, nutritious food" — a number of critical questions come to the fore. Are we dealing "simply" with an issue of poverty? Or is the issue access to food and its delivery, or is it a matter of food production? To put it another way, should we be worried about the world's ability to grow enough food for its entire population or do we "simply" have a distribution problem?

Obviously, the causes of hunger are both complex and heavily intertwined with hosts of other factors:

- *Poverty:* Hunger and poverty are inextricably inter-twined, with one feeding (*sic*) off the other. The role of poverty in hunger becomes, of course, more critical in parts of the world where food is scarce.
- *World population:* As the number of mouths to be fed increases, so does the need to produce more food and to maintain standards of living to avoid poverty and the associated hunger. Not surprisingly, the world's increasing population is a major driver of all the issues in food security and environmental change. Urbanization and the growth of the middle class (e.g., tens of millions of people per year moving from rural subsistence living to living middle class lives in China and India) represent a major challenge to food producers and distributors.
- *Food and agricultural policies:* How robust is our agricultural productivity? How good are our harvests? How nutritious are the foods that we produce? How good is our ability to adapt to changes in food require-ments and how effective are our systems to transport food? In all areas of the developed world, and increas-ingly in underdeveloped areas, governments are playing important roles in food security.
- *Climate change:* There are many who will claim that changes in climate are already the biggest threat to food security and a cause of hunger and poverty. Changes in temperature and rainfall; dramatic weather events that bring about extremes of drought, heat, flooding and fires; soil erosion and contamination of coastal waters all have enormous impact on food security.

Many experts believe that between population growth, the growth of urban areas and changes to the environment, we do have an impending problem in terms of producing enough food to feed the world. It is therefore critical to understand the differences between populations starving because of lack of access to food and the issue of severe global food shortages.

A HISTORY OF HUNGER

Thousands of years ago, the Egyptians, Romans and Babylonians were also concerned about the security of their food supply and routinely used minerals and or manure to enhance the productivity of their farms. Not dissimilar to our own current situation, the availability of water also made productivity a problem.

By the time of the first industrial revolution, particularly in Europe, people had started to move from rural to urban settings. Attention had to turn to the nature of food production and delivery. The possibility of food shortages and the role of fertilizers in food production started to draw a lot of attention. As early as 1840, scientists discovered that the key ingredient in fertilizer was nitrogen. However, while there is abundant nitrogen in the atmosphere — it makes up more than 80 percent of the air we breathe — it isn't in a form that can be used for growing most crops.

Skip forward to the late 1880s, when the German government foresaw a crisis looming. The nation's leaders, aided by scientific evidence, determined that while they had the land and sunlight to feed upward of 30 million people, without the necessary fertilizers to increase yields, as many as 20 million people would starve. As a result, an important industry emerged: the importation of nitrogen-rich fertilizer (mostly seabird guano) from caves in Chile and Peru. The supply and cost were far from stable, so the government invested in research to find ways to obtain new sources of nitrogen-based fertilizers, and in the early 1900s, a brilliant young German chemist by the name of Fritz Haber had discovered a chemical process (the Haber–Bosch reaction) to produce ammonia from nitrogen and hydrogen — a form of nitrogen that is readily available to stimulate plant growth. This reaction requires significant energy to run, most often in the form of electricity.

Today, the process produces one hundred million tons of synthetic fertilizer annually to help feed five to six billion people. It is estimated that half of the nitrogen in our bodies is a result of Haber's discovery of ammonia-based fertilizers. Strikingly, scientists have calculated that without that discovery, the human race would have topped out at 1.5 billion people, and that any population growth above that would have resulted in massive starvation. I like this fact because it adds credence to the importance of research and public policy in something as simple as putting enough food on our plates and sustaining population growth. And we might think that today we sit in a similar state, but with a worldwide population of 7.6 billion people, it certainly behooves us to ask if we have the wherewithal to feed the 9.8 billion people that we expect on Earth by 2050?

Let's skip forward to the second half of the 20th century. Prior to 1960, when the world population was approximately three billion, there was plenty of talk about starvation. Discussions largely centered on the issue of access to food. Overpopulation in certain regions, high birth rates, poverty and climates that did not easily support farming (such as sub-Saharan Africa) led to massive malnutrition and high death rates. There was, however, little talk of global food shortages. This began to change in tandem with a sharp increase in the world's population.

Rachel Carson really gave birth to the entire environmental movement. She was a writer, a scientist and an ecologist working for the US Fish and Wildlife Service (USFWS). In 1962, she wrote one of the most important books of our time, *Silent Spring*, which was to become a critical tipping point for the environmental and conservationist movements. Her powerful work demonstrated the detrimental effects of indiscriminate use of chemical pesticides on the environment, which resulted in a ban (in the United States) of the common pesticide DDT for agricultural uses. This was really the start of the environmental movement and led to the creation of the US Environmental Protection Agency (EPA).

While Carson's work was seminal in terms of environmental activism, it attracted huge criticism. For example, Robert White-Stevens, who worked for the US chemical/agriculture company American Cyanamid, wrote:

> **"If man were to follow the teachings of Miss Carson, we would return to the Dark Ages, and the insects and diseases and vermin would once again inherit the Earth."**

Even in 2019, there are still diehards who resist science and who would attest to Stevens's quote. But Carson's work stuck, and agricultural technologies changed dramatically. However, major concerns surfaced about how a growing population might be fed in the wake of bans on DDT and other environmentally nasty chemical fertilizers. In his best-selling book, *The Population Bomb* (1968), Paul Ehrlich warned about the consequences of continued population growth, especially in the face of limited resources. His prediction was grim, if not alarmist:

> **"The battle to feed all of humanity is over . . . in the 1970s and 1980s, hundreds of millions of people will starve to death in spite of any crash programs embarked upon now . . . I have yet to meet anyone familiar with the situation who thinks that India will be self-sufficient in food by 1971. And indeed, India couldn't possibly feed 200 million people by the year 1980."**

Ehrlich was of course criticized for hysteria and for his doomsday approach, especially when his predictions failed, at least in the time frame he suggested. He was viewed as an alarmist and called an "irrepressible doomster" (which he was). Nevertheless, he continued to write about how the world's growing population was far exceeding Earth's capacity to sustain current Western living standards.

While Ehrlich may have been off on his time frame, his warnings about population growth and its pressure on food and other resources were clear. Many concluded, of course, that Ehrlich was just dead wrong and a kook. But clearly, it's not so simple. The huge population growth has put major pressure on issues related to our food supply as well as issues pertaining to our environment. Fortunately, people and especially governments started to take notice.

During roughly the same period, Dr. Norman Borlaug was hard at work on a solution. Dr. Borlaug was a prominent American agronomist who received the Nobel Peace Prize in 1970 in recognition of his contributions to world peace through his work in increasing the global food supply. As a young man in the mid-1930s, Borlaug worked with the unemployed on federal conservation projects. He saw firsthand the impact of starvation and the critical importance of an adequate food supply in general and for the workforce in particular. In 1942, Borlaug took up an agricultural research position in Mexico, where in a few short years he developed several varieties of semi-dwarf wheat that had a high yield and were disease resistant. In the 1950s, he led the introduction of these high-yield varieties and other modern agricultural production techniques in Mexico, Pakistan and India. Within just a few years, the impact was enormous. By 1963, Mexico, even with high birth rates and significant population growth, had become a net exporter of wheat, and by 1970 wheat production had almost doubled in both Pakistan and India. Longer term, between 1960 and 2015, wheat yields in some of the least-developed countries of the world (but not in many parts of Africa) more than doubled from less than 1,000 kilograms per hectare to more than 2,100 kilograms. As I mentioned earlier, Dr. Borlaug became known as "the man who saved a billion lives" because his work developing these new strains of wheat meant that Earth's food supply could now sustain an additional one billion people.

Though he never used the modern tools of biotechnology or genetic engineering in his work, Borlaug felt strongly that, given the rapidly increasing need for more food production and the limits to expanding the amount of arable land, genetically modified foods (currently called GMOs) would be critical in achieving the levels of food production necessary to feed the world.

Rachel Carson and Norman Borlaug were giants of the 20th century who did more than any other scientists to further agricultural productivity. While Carson didn't coin the term (and never even used it), her work really defined the issue of *sustainability*. That's a critical concept as we attempt to figure out how to feed the world in the face of a rapidly increasing population, major urbanization and a variety of environmental pressures. Many of these latter pressures — destruction of forests, increased burning of fossil fuels, increased use of chemical fertilizers and so on — emanate from the increased requirement for global food production and, of course, all are critical with respect to issues of greenhouse gases and global warming.

WHAT'S ON THE MENU?

So, where do we stand currently and what's on the menu for the coming decades? In 2014, Fred Davies, a top scientist at the US Agency for International Development, announced:

> "For the first time in human history, food production will be limited on a global scale by the availability of land, water and energy. The increases currently projected for crop production from biotechnology, genetics, agronomics and horticulture will not be sufficient to meet food demands."

I underscore the issues of land, water and energy availability because these are relatively new considerations that emerged toward the end of the 20th century. It is evident that the increased access to these commodities will become an important limiting factor as we deal with continued population growth, industrialization and the challenge of environmental issues.

Is Davies right to be worried? Are we producing enough food for 2021 or 2050? There are too many variables to answer that question with any degree of certainty. As Niels Bohr, a Nobel laureate in physics, once said, "Prediction is very difficult, especially if it's about the future." Nonetheless, in Figure 16 I attempt to summarize the critical issues with regard to the adequacy of the world's food supply.

Figure 16: The Perils in Our Food Supply

1.	Population growth, especially in Asia and Sub-Saharan Africa
2.	Yields of agricultural crops have begun to plateau
3.	Agriculture's impact on the global enviroment continues to grow, producting 24% of greenhouse gas emissions, using 37% of the earth's landmass and using 70% of its fresh water
4.	Disproportionate growth of urban population in developing countries
5.	Global warming changes growth paterns often, producing severe weather, drought, fire, flooding, etc.
6.	Energy costs are increasing, driven at least partly by the many demands of agriculture
7.	Clean water sources are scarcer and more expensive
8.	Global demand for protein from meat and milk products are increasing
9.	There is a general consensus that the Food Gap (69% increase in food calories between 2010 and 2050) cannot be closed using current technologies and growth conditions
10.	Major news approaches/technologies are critical

The question of food is really a question about the future of mankind and when (and some might still say if) a real-world food crisis will actually occur — or is the crisis already here and just not yet recognized? One of the challenges we have in the Western world is recognizing the problem in the face of plenty. The fact is that the security of the world's supply of food is in real danger, given the continued growth of the world population, huge increases in urbanization and, equally critical, major changes to the environment, including but not limited to climate change.

Faced with those growing threats, the agriculture industry and its various regulators and policy makers have their plates full of critical questions, including:

- How large will the population grow?
- Will the rate of urbanization continue to grow?
- Will population growth in poverty-stricken areas of the world continue unabated?
- How much food are we producing today, and what are the growth prospects in terms of technologies and conditions for a continued increase in food production?
- What new technologies can be brought to bear to increase food production?
- What are the issues around land use and the potential to increase the amount of arable land?
- What is the implication of continuing to cut down forests in favor of agricultural lands?
- What is the impact of climate change on food production, including the amount of arable land available?
- What is the impact of agriculture on greenhouse gas production and global warming?
- How do we factor in water and energy use?
- How do we deal with the public's concerns about GMOs and other new technologies?

We're not going to address all these issues in this book, but I would argue very strongly that they are *all* of critical importance and that each has to be dealt with. Furthermore, unless we figure out how to increase food production and do it in a sustainable way (that is, without continuing to devastate our environment), we will indeed face a major gap, the difference between what we need and what we produce, and it will come well before 2050.

GMO SOLUTIONS

Before we move into potential solutions that synthetic biology can offer, it's worth examining the advances that have been made with GMOs, as they can be considered a precursor to synthetic biology solutions. Since the introduction of the world's first GMO, the Flavr Savr tomato, more than 25 years ago, scientists have been developing GMOs to address our global food-related issues. In fact, you may be surprised to learn that in the Western world, large quantities of food crops (sometimes more than 80 percent) — including corn, canola, soybeans, sugar beets, alfalfa, yellow squash, zucchini, Hawaiian papaya and cotton — are already predominantly GMOs.

So, what exactly is a GMO? There are many definitions, and two of them are as follows. The Institute for Responsible Technology considers GMOs to be "a result of a laboratory process where genes or DNA from one species are extracted and incorporated into the genes of an unrelated plant or animal." A second definition says that a GMO is "an organism whose genome has been altered by the techniques of genetic engineering, so that its DNA contains one or more genes not normally found there."

The first definition deals with mixing genes from unrelated species, whereas the second definition inserts the words "genes not normally found." That definition, as it turns out, is somewhat arbitrary, since through evolution — and even more so through intentional crossbreeding, which has now been used for centuries

— genes appear and disappear within specific organisms. What if a gene exists in a species but is never expressed and then you manipulate it to express itself? Is the result then a GMO or not? The new techniques of gene editing further complicate the definition.

I personally don't get too fussed with the definitions. What's more important is that we have a mandate — really an obligation — to make sure that we have enough safe and healthy food to feed the world's increasing population, and techniques of genetic modification are critical to this objective. This again is not new; crossbreeding and selective hybridization are genetic tools that have been used for centuries, and they were typically developed to provide the population with more and better crops. We now simply have more sophisticated and more specific tools to create new foods that are so critical for our continued survival. The issue of safety, which will be discussed, is, to my mind, much more important than attempting to make quite arbitrary distinctions.

As with the introduction of any other new technology, you would expect some of the GMO products to have overwhelmingly important implications and some to be somewhat more trivial. Some examples follow in the paragraphs below.

In the majority of instances to date, GMO crops have very important advantages over their non-GMOs counterparts, as the first objective has been to make the crops resistant to pests and weeds. That is, a gene is introduced into the plant crop which makes it resistant to pests so that pesticides are no longer required in the crop's growth. The farmer is delighted, as should be the consumer, because there is less need to apply pesticides and herbicides to the crops, thus saving money for the farmer, increasing yields and potentially increasing the health benefits to the consumer.

A different example of a GMO is an attempt to incorporate an HIV-neutralizing protein into the genome of rice to try to minimize the spread of HIV/AIDS in underdeveloped countries. This produces a GMO crop that has the potential to deploy a durable and inexpensive anti-HIV agent throughout the developing world. This is an excellent example of how new technologies can be employed at relatively low cost with incredible health implications especially in developing countries.

Another, albeit less important, example is an apple called the Arctic, a product of genetic engineering (a GMO) that is being harvested in Washington State and has been approved for sale by the American Food and Drug Administration (FDA). The apple is "non-browning," so it remains fresh even after you've taken a bite out of it or left your cut-up apple quarters on the table. This was achieved by "silencing" the gene responsible for browning. You might ask why the world needs a non-browning apple; doesn't that just offer a simple cosmetic change? In fact, a non-bruising, non-browning apple does two things, both important from the consumer's point of view and one important from a food management point of view. There's the straight commercial benefit that the apples are attractive to eat and therefore more people will eat them. But perhaps more important in terms of food security and environmental benefits is that there will be much less wastage, as slightly bruised apples do not end up in the trash heap or composting bin.

If we adhere to the idea that there is indeed a major threat to the adequacy of the world's food supply, we can very quickly conclude that a host of technologies — principal among them, genomic technologies — will be integral to solving the issue of food security. As such, GMOs will likely continue to be developed and, in fact, become the norm. However, selling GMOs (and synthetic biology) as a major solution to the world's food problem poses some real and very interesting issues with regard to the future of food, even including disbelief, as in: *Is this really a problem?*

On the issue of food security, we have an important dichotomy. On the one hand, there are those of us who are relatively well-heeled and living in the developed world, where not only is food plentiful and often relatively cheap, but there is an increasing demand for fresh, natural foods that are clearly labeled. Organic foods are considered healthier, and outlets such as Whole Foods and the Food Emporium do a bustling business. Processed foods are frowned upon, even though the definition of "processed" is not clear; additives are considered unhealthy or even toxic, despite the lack of strong evidence to support that belief. Some people in the health food business go even further, insisting that "purity" must be maintained in the growth conditions as well — in order to provide "healthy" produce, farmers must avoid using chemical fertilizers on their land and avoid providing antibiotics to their animals to increase their productivity (and, of course, their profit). Perhaps more than anything, there is mistrust on the part of the "educated" consumer toward the multinational food companies, who are perceived to be willing to do anything to protect their profit margins. This is largely just unfounded rhetoric. But I digress. On the other hand, there is much of the rest of the world who are happy to get just a meal on the table (often, of course, there is no table) and a sufficient amount of calories in their bellies.

Since the introduction of the the Flavr Savr tomato (and in spite of the ranting and raving of the anti-GMO crowd, which I discuss in more depth in chapter 8), two very clear conclusions can be made. First, throughout countless studies, GMOs have been proven safe for human consumption, with no documented harm to any other life-forms. In fact, there has been not a single legitimate documented case of the consumption of a GMO doing any harm to any human being or animal. Second, GMOs have repeatedly been shown to be beneficial to corporations and farmers and society: yields have increased; pesticide use, deforestation and waste have all decreased; and crops have become more profitable. As you will note from the list below, many of these benefits have accrued not only to Western agriculture but also to poorer areas of the world.

- Yields of GM corn in the Philippines have increased household incomes by more than 50 percent.

- In Burkina Faso, profits have increased from $70 per hectare to $150 since the introduction of GM cotton.
- Chinese producers of GM cotton have realized a 71-percent decrease in insecticide use compared to non-GM cotton.
- Farmers in Western Canada realized a $350–$400 million increase in revenue as a result of planting GM canola.
- An overall analysis of 147 studies of GM crops described a 37 percent decrease in pesticide use, a 22 percent increase in farm yields and a 68 percent increase in farmer profits. It seems to me that that's not a bad thing. Yes, corporations are profiting, but we're also producing more and better crops at a time where the security of the world's food supply is reported to be in some peril.
- An environmental impact study by Brookes and Barfoot has shown that the carbon dioxide emission reductions brought about by growing GM crops were equal to removing ten million cars from the road for one year.

This last point is quite extraordinary in terms of the relationship of agriculture to environmental concerns, and it might even represent the most important benefit of GMOs.

SERVING UP SYNTHETIC BIOLOGY

There are a number of important questions to consider when it comes to the security of our food supply, and the answers will no doubt dictate the rate at which synthetic biology processes will become a major source of our food.

- What will the world population be in 2050?

- Will population growth in urban areas in countries such as India, China and parts of Africa continue at the current rate?
- Will Earth's temperature continue to rise?
- Will the oceans continue to rise, causing flooding in coastal regions?
- Will climate change produce extremes of drought and flooding in food-growing areas?
- As a result of global warming, will we be growing more food farther north in the northern hemisphere and farther south in the southern hemisphere?
- How much farther can we go to produce better farm yields with traditional technology?
- Will the costs of water and energy for agriculture become prohibitive?
- Will the cost of mined and chemically synthesized fertilizers continue to rise?
- Will the impact of agriculture on the increase in greenhouse gases and resultant global warming become so severe that drastic changes in the way we grow food become an imperative and maybe even legislated?

These are all critical questions, each of which has the potential for massive impact on the security of the world's food supply. We have no option other than to address each of them if we are to prepare for ensuring an adequate food supply.

As we discussed earlier, CRISPR-Cas9 is becoming an important tool for scientists who wish to edit genes to combat cancer, to make plants drought-resistant or to modify bacteria to bind mercury or lead or clean up oil spills. Note that once again we're taking a lesson from biology. The CRISPR system is able to identify a particular gene sequence, snip it out and replace it with a more appropriate bit. It can be used to either snip out a bad (i.e., mutated) gene sequence and replace it with a good one or even to add a new sequence. This is a much easier and better method than many of the gene-splicing technologies developed over the past two to three decades. And why is this one better? Likely because it's a natural biological process: it was developed and improved over many millions (or even billions) of years to protect the well-being of bacteria and fend off attackers.

But this leads to a question that has begun to cause rumination among scientists, regulators and those who oppose the progress associated with genetic modification: Is a food or a plant that has had its genes edited using CRISPR-Cas9 a GMO or not? The jury seems still to be out, but to my mind this is just a matter of semantics. We know that society takes on new technologies readily when it is in their interest, and GMOs and the technologies that produce them are in that category. It is simply incumbent upon us to regulate the technology and to determine its utility and safety.

So, what is the difference between GMO foods and foods offered by synthetic biology? In general, GMO scientists take genes with specific desired characteristics from one organism and cut and paste them into the genome of another. The result could be a simple hybrid, or you might take a gene for an "antifreeze" protein from a fish that lives in frigid deep-sea conditions and implant it into the genome of a grain crop to make it resistant to frost or cold. Synthetic biology, on the other hand, starts with genes from any source that are stored in computer code and then remixes those genes. This is first done in code on the computer and then by synthesizing the DNA and having it expressed in cells, creating food entities that may be entirely novel. This process is very succinctly described by Josie Garthwaite, a prominent science writer, in an *Atlantic* magazine article about the future of food:

"If genetic sequencing is about reading DNA and genetic engineering as we know it is about copying,

cutting and pasting it, synthetic biology is about
writing and programming new DNA with two main
goals: create genetic machines from scratch and
gain new insight about how life works."

The new "genetic machine" might be a new plant or food that contains all the ingredients coded for by its newly made DNA. That genetic machine might code for, and therefore produce, an entirely new vegetable that has the combined taste and growth characteristics of a tomato, a carrot and a potato with growth conditions that require less water, less sunlight and lower temperatures.

The Americans, principally in California, and the Israelis have become especially adept at growing food with minimal water waste, using techniques such as drip irrigation, and they have some of the most aggressive food research programs. An interesting thing about farmers, and this is probably true throughout the organized or commercial farming world, is that they have traditionally been very good at selecting particular crops and adapting new technologies to increase total food yields as a function of growth conditions, price and demand. This probably bodes well for the pace at which new food technologies are adapted, the critical caveat being consumers' understanding and acceptance.

But experts insist that there are limits to increased productivity using traditional means. If climate change demands that we produce crops that are drought-resistant or that require different temperatures or daylight cycles as agriculture moves farther from the equator with temperature increases, then those changes will have to take place. And what about the coastal regions of Africa and Asia where seasonal flooding with seawater results in brackish (salt-contaminated) water that is unable to support traditional agriculture? Will those salt-contaminated fields merely lay fallow, or will we be able to modify existing crops or develop entirely new ones to incorporate genes that support plant growth in soil with a high salt content?

The good news is that exciting new technologies in genomics for agriculture, along with increased and more effective productivity, are rapidly being developed. Increasingly, as the genomes of most (if not all) of our foods become known and the roles of specific genes/proteins are determined, we will likely be able to use the tools of genetic engineering and synthetic biology to create crops that will grow in previously inhospitable environments. In far northern (or southern) climates, where the light cycles are different, we will introduce genes that alter the light requirements of the plant. In cold climates, we will introduce antifreeze proteins to protect against early or late seasonal frosts. In more arid climates, we will introduce characteristics that cause plants to use their limited water resources more frugally. And in soil that has a high salt content as a result of coastal flooding, we will introduce genes that allow plants to thrive in conditions of high salinity.

While I've alluded to this several times already, if we grow these foods in large laboratories and not in fields, this development alone will have a major positive impact on the environment. By putting a stop to clear-cutting forests to make way for agricultural land and by reestablishing those forests, we should see a dramatic drop in greenhouse gases and an actual reversal of global warming.

Now for a few modest predictions. It's 2040 and the world's population has continued to grow unabated. (I've used the year 2040 rather than 2050 because I'm concerned that the crises humanity is facing are going to come upon us a lot more quickly than most of us expect.) Major industrialized countries have started to take issues of greenhouse gases and global warming very seriously. Global environmental changes have continued to offer major challenges to agriculture, and the industry is increasingly being required to examine its carbon footprint — its own impact on greenhouse gas production and global warming. Entirely new foods and even new food groups are being produced, much of them grown under vastly new conditions, resulting in less land use, less energy consumption, less water usage and much less pollution. The foods have also been created for better nutrition and taste and are grown more efficiently, with less wastage. The public has adapted quickly to these changes because: a) they've seen the need for change and b) the changes have been made easy in terms of cost and attractiveness, including issues of feel, smell and taste of the new food. We've arrived at a point where we can design foods exactly according to our needs and wants and, perhaps most importantly, with a much-reduced carbon footprint. In developing these foods, synthetic biologists will consider:

- What protein, fat and carbohydrate content and type do we want in our food?
- Are pests an issue? Do we want our foods to be pest-resistant, both during and after cultivation?
- Do we want to grow our food in different light, temperature or moisture conditions?
- Do we want to lower transportation costs and grow/ produce more food locally?
- Do we want to alter the taste, the texture or the color of our food?
- What conditions do we need for food storage?

Again, I stress that what's incredible is that none of this represents a flight of fancy. With our new knowledge of genomics and the power of synthetic biology, achieving these aims is now well within reach. While the earliest food products of synthetic biology are not yet available in the grocery store, they are at least being realized in the laboratory. For example, a number of companies, again most commonly in the United States and Israel, are working on growing beef and chicken in large vats of cultured cells rather than on the bodies of actual animals. And if you're worried about how the product looks, you can be sure that someone will use a digital printer to make your porterhouse steak or chicken wings grown in laboratories look just like the ones Grandma used to make.

You might consider some of these ideas pretty dramatic, or even bizarre, but remember that many of the foods we eat today look very different than they did a hundred years ago. You would not recognize the corn, carrots, eggplant or watermelon that was available in 1900. All of them have been selectively bred to provide more suitable growth and taste characteristics. So, there's no reason to expect that the food we'll be eating in 50 years will be similar to what we have today. Below are some examples of what synthetic biologists are developing.

Waste Not, Want Not

It is estimated that as much as 50 percent, and maybe even more, of food grown is wasted, that it is not consumed but lost to "garbage." This wastage can occur at any of four stages of the food cycle: growth, processing, retail or consumption.

The process of composting food and other organic waste to create more and possibly better fertilizer has received a lot of attention in recent years, especially among those with a well-developed concern for the environment. Overall, however, composting has yet to make even a small dent in addressing the huge issue of food wastage. Far better would be to create a system of food waste management in which, at every level from growers to consumers, the essential nutrients in the discarded food are not only used as fertilizer but are also recycled to produce protein, carbohydrate, fats and all the other nutrients essential to our diet, including vitamins. It doesn't take much imagination at this point to envision large vats of synthetic cells with the cellular machinery necessary to convert all that waste back into nutritious food products or ingredients.

Genecis is a student-run company that recently spun out of the University of Toronto. It essentially uses techniques of synthetic biology, applying advances in biotechnology, microbial engineering and machine learning to take food that would otherwise find its way to landfills to make PHAs (polyhydroxyalkanoates), which is a high-quality and fully biodegradable form of plastic suitable for all sorts of different uses.

But it's not just an issue of recouping calories or decreasing waste. Effective recycling of waste food products will constitute an essential "attack" on the issue of greenhouse gases (GHG) and global warming. In the Unites States alone, food waste produces over one billion tons of CO_2, which is almost as much GHG as is produced by the country's entire transportation sector. This is a stark reminder of the very significant relationship between global warming and the way we feed the world.

SuperMeat

In 2015, SuperMeat, a brash young Israeli company with an exciting fundraising campaign, announced its newest product, SuperMeat — real meat, made without harming animals. The concept is really quite simple, and a number of California companies are also thinking and working along the same lines. Why grow meat on a live animal when you can grow it in culture in a large tank?

Think about it for a moment. You grow massive numbers of beef cells in culture and then you use a digital printer to give the cells the form of a sirloin steak or, if you'd rather, a porterhouse steak or even a hamburger. Not so crazy, and the benefits are fantastic. According to the SuperMeat company (the numbers may be exaggerated, but the concept is clear):

- Growing cultured meat requires 99 percent less land, emits 96 percent less greenhouse gas and uses up 96 percent less water than the traditional meat industry uses today.

- SuperMeat will help solve world hunger, as many fewer resources are required to feed a rapidly increasing demand.
- SuperMeat will be healthier than regular meat because the manufacturing process is much better controlled.
- SuperMeat will cost a fraction of conventionally grown meat. Why? Because culturing meat and mass producing it will be much cheaper than growing and feeding billions of animals.

The cells being used to culture meat can be little more than random cells taken from a beef cow (muscle, liver, kidney or whatever), from a chicken or, for that matter, whatever animal you choose. And you only need to take the cells once, as you allow them to continue to proliferate ad infinitum or as long as you want in a very controlled culture environment. Furthermore, the technologies of synthetic biology can be used to control taste, smell, texture and, of course, nutrition. In fact, taste can be completely controlled and adapted to suit the desires of the consumer's taste buds.

The "yuck factor" disappears when you go to the grocery store and discover a sirloin steak that looks just like a sirloin steak, except it was grown in culture and not on the bones of a cow that had to be slaughtered to get the meat. And note that this is real, high-protein meat, not just a vegetable- or legume-based product manufactured to look and taste like meat.

SuperMeat and similar companies have developed prototype samples of their meat and are now in scale-up mode to perfect the processes and critically, of course, to improve cost-effectiveness. Many hundreds of millions of dollars have already been invested in these companies by venture capital firms.

Milk without Cows

The notion of growing food in vats of individual cultured cells rather than from whole animals has garnered its own name: "cellular agriculture." A California company, Perfect Day, is one of a number of companies using genetically engineered yeast that has been programmed by inserting the appropriate DNA to produce casein, lactoglobulin and lactalbumin, which are the three critical and most abundant proteins found in milk. This creates a "milk" with the exact same proteins that you get from actual cow's milk, but no cows are needed. Furthermore, these same yeast cells can be programmed to include any other components that the producers might want in their product. Think of it for a moment: a whole variety of milks fortified with whatever ingredients you like, and these fortifiers are made with the milk, not included as additives. There are likely no limits to the technology, as long as you understand the genes necessary for the cells to produce the desired ingredient.

Craig Venter says it best:

> "Agriculture as we know it needs to disappear.
> We can design better and healthier proteins
> than we get from nature."

Obviously, Venter didn't mean the mere tweaking of foods to make them healthier, cheaper or better looking. He was looking at wholesale changes to what we eat.

Moving to Mars

Let's examine the potential of synthetic biology to an extreme with respect to the future of our food supply. I don't expect you've thought much about what you're going to eat when you get to Mars, but given the deteriorating conditions here on Earth, others have given it a good deal of thought. Before you dismiss this as a flight of science fantasy, think for a moment about how important the various space exploration initiatives, including Neil Armstrong's walk on the Moon, have been for technology development. So, regardless of whether we actually colonize Mars, the impact of the technologies developed in planning for it will have enormous implications here on Earth. This is especially so in the realm of food production, so let's consider how you might feed yourself when you get to Mars.

You certainly couldn't simply ship food to Mars. It was initially thought that all sorts of seeds could be generated here on Earth to be able to find a few that would be able to grow in the atmosphere of Mars or in enclosed, controlled conditions, protected from the potentially harsh Mars environment. With the advent of synthetic biology, we can shed the notion of only being able to choose to modify known terrestrial foods. If we have the technology, why not design entirely new foods to suit our needs — in this case, to grow on Mars?

Let's start by thinking about how to grow a healthy new vegetable that is 40 percent protein, 40 percent carbohydrate and 20 percent fat, incorporates critical minerals such as calcium and magnesium and produces any number of essential vitamins. Furthermore, it should withstand nightly freezing conditions, require only four hours a day of sunlight, use water very sparingly and withstand the levels of ionizing radiation found on Mars. And, of course, because we're earthlings, we'd want to consider taste, smell, consistency and other issues of palatability.

Designing the perfect food from scratch (in this case, say, a vegetable) would have been pure science fiction a short number of years ago, but today it's a realistic objective. And by the way, while we're at it, why stick just to vegetables? Why not be able to also incorporate animal proteins, for instance, the major proteins that are in cow's milk? Within just a short period of time, many of the genes that confer the above properties on a plant or animal will have been discovered, if they haven't been identified already. The genetic or DNA sequence will be stored in code on the computers that will be delivered (with you) to Mars. So, now all you need to produce these new vegetables are:

- a computer containing the DNA code for all the genes to make all the proteins necessary for the plant, vegetable or whatever you want to grow on Mars;
- bottles of the four base pairs — the four chemical components of DNA (adenine, guanidine, cytidine and thymidine) that are needed to synthesize the DNA/genes;
- a DNA synthesizer to read the computer code and make the genes from the contents of the four bottles;
- a starter cell line (just a few), similar to the empty bacterium that Craig Venter used to make the first line of synthesized "live" cells; and
- the nutrients necessary for plant growth — carbon dioxide, a host of basic minerals such as sodium, potassium, nitrates, phosphates and sulfates, and a number of more minor ingredients. Some of them will already be available on Mars, some may be substituted for and some, hopefully very few, may have to arrive with you on your spaceship.

Whether this new food will be grown in Martian soil or in large vats of liquid will require further exploration of the growth conditions on Mars, but those are surely mere details. Maybe we won't go to inhabit Mars, maybe we'll just stay here at home. We can still utilize these same technologies to make new foods here on Earth that are not only tasty, but that also go a long way to closing the food gap, making our food supply more secure. In addition, they will be grown in conditions that are very environmentally friendly;

no more cutting down trees to plant vegetables. Vegetables and their carbon footprint (excess GHG) can be reduced and maybe even eliminated.

A Less-Appetizing Alternate Food Source

Since I've referred to science fiction a number of times, just for fun let's take a moment to look at an alternative that was offered some 45 years ago, albeit in a fanciful storyline. We've been pondering the issue of our food supply in light of serious climate change for many years. In 1973, Charlton Heston and Leigh Taylor-Young, along with Edward G. Robinson (it was to be his last film), starred in an American science fiction thriller called *Soylent Green*. In French, the title was more aptly *Soleil Vert*, or green sun. It takes place in 2022 in New York City, where 40 million people live in horrific dilapidated conditions. The 20th century's industrialization has caused massive overpopulation and pollution; smog permanently hides the sun, turning the air and sky greenish. The population is surviving on a product called Soylent Green, produced by the Soylent Corporation. The inhabitants are told that the green wafer is derived from high-energy plankton from the ocean. But, in a tale of murder and mystery and good detective work, it's clear that the oceans are in fact dead and no longer produce plankton. So, where is the stuff coming from? A *New York Times* reporter deduces that Soylent Green has a protein composition that could only come from human remains. In an intense search for the source, a New York City detective discovers huge garbage trucks scooping up elderly or derelict people in the middle of the night and taking them to Soylent's central factory to convert them into food. Totally gruesome.

I'm recounting this film not to alarm you, but as an indication of my optimism by contrasting the exciting science-based solutions that exist today with problems that existed — and were sources of nightmares and dystopian thinking — not that many years ago.

Making Other Products from Agriculture

Just as we've learned from biology, we are also learning from agriculture (plants and animals) to make better products, in this case products other than foods. Professor Oded Shoseyov of the Hebrew University of Jerusalem works in the areas of plant biotechnology and nanomaterials. Although Shoseyov doesn't identify himself specifically as a synthetic biologist, his work and his ideas are indicative of the power of these new technologies. He offered a wonderful quote in a TED Talk:

> **"If you want to have a new idea, open an old book!**
> **The book was written over three billion years of**
> **evolution, and the text is the DNA of all living**
> **organisms. All we need is to read the DNA code**
> **and start our progress from there."**

In his talk, Shoseyov makes the point that two hundred years of modern chemistry have managed to create synthetic materials that do not come close to many of the biomaterials that nature developed over three billion years of evolution. For example, he and his colleagues combined the strength of crystalline nanocellulose from plants — weight for weight, as much as 10 times stronger than steel — with the elastic properties of a protein called resilin from cat fleas (it gives them their astounding jumping ability). These genes combined to make a very strong, flexible material for any number of different applications. It is incredibly tough, transparent and sufficiently flexible to be used in products ranging from transparent films to biomaterials such as artificial tendons (used, for example, in Achilles tendon replacements). Heart valves made with the material are 350 percent tougher and 300 percent more elastic than the originals! You need only use your imagination to consider the possibilities.

AGRICULTURE AND CLIMATE CHANGE

When we think of environmental damage, we think about cars and factories burning fossil fuels and generating huge amounts of greenhouse gases, leading to global warming. And we think about industries, including mining companies spewing out a range of pollutants that contaminate the air we breathe and the water we drink. Something that we often don't think about is just how intensive the energy requirements and environmental impacts are of virtually all aspects of our food production: from growth of plants and animals on the land, all the way to dealing with waste. For example, water is an increasingly rare commodity and agriculture is currently responsible for as much as 70 percent of water usage. We often think about our cars and our frivolous use of disposable plastics as the major source of greenhouse gases and general pollution and believe that if we could just limit our use of our cars or go electric and recycle everything, life would be good and we would be on our way to saving the planet. Not so fast.

Agriculture is both a source and a sink for CO_2, the major culprit in greenhouse gas emissions and global warming. Plants and trees use CO_2 from the atmosphere. If it were just that, agriculture would lead to a decrease in atmospheric CO_2 and a decrease in greenhouse gases and global warming. Unfortunately, agriculture is also a major producer of CO_2, from its use of energy and burning of fossil fuels, to the cutting down of our forests to create more agricultural land, to the need to use chemical fertilizers — all result in an ever-increasing net production of CO_2, and at rates at least five times higher than the CO_2 produced by our transportation systems.

Consider the demand for protein. Here in the Western world, we have the luxury of being vegetarian or even vegan. Our food stores, and not only Whole Foods, are loaded with different grains and nuts with high protein content. But that's just a small fraction of the world. In most areas of the underdeveloped world and large parts of the developed world, the demand for high-protein food is substantial and the supply is usually poor. The demands for milk and meat at acceptable prices is a major and critical issue in many parts of the developing world.

The problems are compounded as major population growth necessitates cutting down forests to provide room for housing and the land required for farming and raising livestock to feed the expanding population. Everything we do here exacerbates the issue of global warming, as cutting down trees results in less CO_2 being removed from the atmosphere and the demand for more water and more energy to sustain the increased farming all result in increases in greenhouse gases, impacting global warming. As an example, between 1960 and 2010 the worldwide demand for protein increased by 40 percent, with the developed world consuming 12 percent more and the developing world a whopping 60 percent more.

Another example of agriculture's strain on the environment is the industry's reliance on nitrogen. Although a host of other energy-dependent components, such as water, light and heat, are also critical to agriculture, nitrogen fertilizers are the most essential in terms of productivity and yield. What do we know about nitrogen in the context of food and agriculture? Although it is the largest component of the air we breathe (about 80 percent), plants cannot extract nitrogen directly from the air. The problem is that nitrogen gas (N_2) by itself is unreactive, as the bond between its two atoms is very strong. Only in the form of nitrates in the soil can it be taken up and used by the plant. Soil bacteria that fix nitrogen are often the most common requirement for plant growth, but there are problems with all commercial sources of nitrogen fertilizer:

- They are expensive to synthetize and/or mine.
- They use large amounts of energy (coal, gas and electricity), leading to increases in greenhouse gases and therefore global warming and climate change.

- The fertilization process is inefficient: most of the nitrogen reagents are released into the environment, polluting fields, rivers and lakes.

While an easy solution might seem to be growing our food "au naturel," that is, without fertilizers, it's critical to understand that the notion of having the entire world "go organic" is not only impossible, but it would be an environmental disaster. Organic farming requires as much as 50 percent more land use per calorie than conventional farming. We would have to cut down more trees to plant more acreage, and this would only exacerbate the greenhouse gas conundrum. It may be a nice idea, eating healthier and so on, but it's a luxury that only very few can afford and it is environmentally unfriendly. This explains the ongoing search for better, more efficient and less energy-dependent sources of nitrogen fertilizers.

Gingko Bioworks is a Boston-based synthetic biology company that thinks it has a solution. It recently partnered with Bayer AG, a massive German multinational (everything from soup to nuts, from pharmaceuticals to agricultural chemicals). Gingko Bioworks describes itself as an "organism company" that designs custom microbes for a wide range of industries. Through its partnership with Bayer, it intends to disrupt the entire field of nitrogen fertilizers. The objective is nothing short of replacing mined or chemically synthesized fertilizers with microbes, developed through the techniques of synthetic biology. These will pick up nitrogen directly from the air and feed it directly to the plant, thus negating the need for expensive and polluting nitrogen-based fertilizers.

The partners are not dreaming up entirely novel or radical solutions. They are once again taking a lesson from biology — in this case, from the plant microbiome, the hundreds of billions of different microbes in and on Earth that have learned to live in coexistence with hundreds of millions of different plant forms. Microbes associated with delivering nitrogen to satisfy the needs of a variety of plants, such as soybeans, lentils and other legumes already exist. But most crops — especially wheat and other grains — do not have such accommodating microbes. Therefore, they need nitrogen-based fertilizers provided by us to produce the types of yields that humans require. But if Gingko Bioworks' plan comes to fruition, instead of spraying the fields with toxic pesticides, we'll be spraying them with a fine mist of specially synthesized microbes whose primary and even sole function will be to grab nitrogen from the air and give it directly to the plant. Stay tuned, the work is already in progress.

While serious change to human behavior often comes slowly, the need for change in agriculture is obvious with (or even without) the apparent need to increase food production in the decades before us. Extensive monoculture farming, heavy pesticide and antibiotic use and factory-farmed livestock are incredibly disruptive to both human and animal health and the health of our environment in terms of energy use, pollution and dealing with waste. But nevertheless, they are essential components in feeding the world's increasing and complex population. Josie Garthwaite, in *The Atlantic* says it most succinctly:

> **"In the face of energy and water constraints, a squeeze on cultivable land, and an imperative to limit greenhouse gas emissions, Synbio (synthetic biology) could also transform the way we farm and eat."**

Garthwaite goes on to define the role of synthetic biology in solving the problems of food production in the future:

> **"By assembling biological systems from the genetic codes catalogued in online databases and fine-tuned through computer modeling, they could deliver more-nutritious crops that thrive with less water, land, and energy, and fewer chemical inputs, in more**

variable climates and on lands that otherwise could not support intensive farming."

And that, of course, is how we're going to colonize Mars or the Moon and, of course, secure the future of food on Earth.

The possibility of using synthetic biology to dramatically alter the entire food industry also has enormous potential to solve many of the environmental issues that we are increasingly facing as a result of population growth, urbanization and a desperate need to expand the scope of agriculture. I can only conclude this chapter with a quote attributed to Hippocrates:

"Let food be thy medicine and medicine be thy food."

Chapter Seven

POLLUTED

Throughout the late 19th and early 20th centuries, it seemed as if nature had the capacity to deal with the pressures that humans were beginning to put on it in terms of significant population growth and industrialization. A rather trite and somewhat cynical statement, "The best solution to pollution is dilution," for a time had some truth to it. Earth seems to have a sort of buffering capacity to deal with some of the insults that we have continued to inflict on it, and for a while, even after the first industrial revolution took hold, all seemed well. But when we started to exceed that buffering capacity, dilution ceased to work and our affronts to the environment begin to accumulate, such that by the end of the 20th century, the effects were not only becoming very evident, they were also starting to have important impacts on our lives. For the most obvious impact, just look at the quality of the air and the quality of the water surrounding us. And if you're not quite convinced of the gravity of the pollution we've created, I would refer you to documentary evidence presented in *The Anthropocene: The Human Epoch* by Edward Burtynsky. You will have great difficulty coming away from that film without great concern for our future.

While it is obvious that the keys to saving our environment are to stop the continuous assaults on it and to attempt to remediate the damages we've already inflicted, we have yet to achieve meaningful results through traditional approaches. Part of this failure can be attributed simply to cost, not only to our pocketbooks but equally to the way we chose to lead our lives. But it's also possible that too many people simply don't understand — or don't care to understand — the implications of our actions. As a society we continue our search for cleaner sources of energy as well as more efficient and cleaner mining and farming techniques. But while that search goes on, we exacerbate the situation with continued growth and industrialization. The Conserve Energy Future website provides an extensive list of the pressures on our environment (which could easily be expanded), including:

Figure 17: Stressing the Environment

Pollution	Global warming and climate change
Overpopulation	Natural resource depletion
Waste disposal	Loss of biodiversity
Ocean acidification	Deforestation
Acid rain	Water pollution
Urban sprawl	Public health

It is clear that with population growth and industrialization over the last century, the overproduction and release of carbon dioxide into the atmosphere as a result of the burning of fossil fuels and resulting in global warming is the most serious culprit. It is, however, by no means unique. Over-farming has resulted in excessive phosphates and nitrates pouring into our lakes and rivers. Contaminated mining sites result in the leakage of highly toxic heavy metals, causing molecules such as mercury, lead and cadmium to leach into surrounding watersheds and water supplies. And, of course, with increased quantities of oil and gas being transported large distances over water and land, serious spills and contamination wreak havoc, contaminating waterways and having a devastating impact on commercial fisheries. Currently, five of the most serious environmental challenges or pressures are:

- fossil fuel consumption and the emission of excessive greenhouse gases
- agriculture's side effects: deforestation and excessive use of land, water, fertilizers and energy
- industrial pollution of our lakes and rivers by mining, agriculture, factories and urbanization
- population growth and urban sprawl
- garbage disposal

The challenges appear so overwhelming that they almost make you want to bury your head in the sand, which of course many people do. But while that may be a common response, it's hardly a viable solution. Before we look into how synthetic biology offers hope for the resolution of some of these challenges, let's examine two of the most serious side effects of the pressures listed above: climate change and contaminated water.

CLIMATE CHANGE/GLOBAL WARMING

To begin, let's attempt to clarify some terms: weather, climate, global warming and climate change. I depend on the Earth Science Communications Team at NASA's Jet Propulsion Laboratory at the California Institute of Technology for many of these terms/concepts.

- **Weather** refers to the atmospheric conditions that occur locally over a short period of time, from minutes to hours to days, as in Mark Twain's famous quote, "If you don't like the weather in New England, just wait a few minutes."
- **Climate** refers to the long-term regional or even global average of temperature, humidity and rainfall patterns over seasons, years or even decades.

- **Global warming**, as the term is commonly used, refers to the long-term warming of the planet since the early 20th century and, most notably, since the late 1970s due to the increase in fossil fuel emissions since the first industrial revolution.
- **Climate change** refers to a broad range of global phenomena created predominantly by the burning of fossil fuels, which add heat-trapping gases (carbon dioxide being the most important) to Earth's atmosphere. These include increased temperature trends described by global warming; rising sea levels; ice-mass loss in Greenland, Antarctica, the Arctic and mountain glaciers worldwide; shifts in flower/plant blooming and extreme weather events.

I'm going to go a bit outside of common convention and use the terms "global warming" and "climate change" interchangeably, since they have so much in common. So please don't infer anything when I use one term or the other. In fact, if you think about it, global warming is really a subset of climate change.

That Earth's climate has changed (indeed, numerous times) over recorded time is irrefutable. For example, it has moved through various ice ages, which occurred from as long ago as 2.5 billion years to as little as ten thousand years. But as the definition above states, the past hundred years, and especially the past 50 years, represent a period of intense global warming that without question results from population growth, industrialization and the associated excessive burning of fossil fuels.

While I choose not to take on the conspiracy theorists and climate change deniers in this section, I want to comment on some of the public's skepticism about climate change and particularly about global warming. I believe that much of it comes from the public's comprehension (or lack thereof) of the concepts of probability and statistics.

Anecdotal statements are not very helpful. They range from "I can't remember when we've had such a warm summer" and "There have been more heat alerts this summer than any time in the past 25 years" to "I've never been so cold in my life" and "Look at all the snow out there — how can you possibly believe in global warming?" People will suggest that climate change, in this case they're almost certainly referring to global warming, must be responsible for the increased incidence of earthquakes, fires, floods, hurricanes and even stock market fluctuations. These may well be manifestations of global warming, but we cannot and must not define its absence or presence in such simple terms.

Which brings me back to the issues of data and evidence and the public's comprehension, or lack thereof, of probability and statistics. In *Mark Twain's Own Autobiography: The Chapters from the North American Review*, the American humorist wrote: "*Figures often beguile me, particularly when I have the arranging of them myself; in which case the remark attributed to Disraeli would often apply with justice and force: 'There are three kinds of lies: lies, damned lies and statistics.'*" (Who actually made that remark about statistics is not clear, but nowhere is it to be found in Disraeli's writings.) My point is that we have to develop and substantiate our arguments with solid data and statistics. I began writing this book in the days immediately after the fall hurricane season of 2018 and the epic storms that devastated huge parts of the Caribbean, Florida and the Carolinas. It would be simple to blame each storm on climate change. However, the fact is that, statistically, the increase in extreme weather seen over the past dozen years or so is distinct compared to storm activity seen in the 40 or 50 years earlier.

The basic sentiment coming from the climate change deniers is: *I don't believe in global warming and I certainly don't believe that, even if it does exist, it is man-made.* They are simply wrong.

Global warming can be attributed directly to the data described in Chapter 3: Man-made. Growth of the world's population began to accelerate in the post–World War II years, around 1950, and this growth was accompanied by huge increases in industrialization, fossil fuel consumption and deforestation. The direct results were dramatic changes to the levels of carbon dioxide and greenhouse gases in the atmosphere, acidification of our oceans, melting of the polar caps, extreme urbanization and so forth. Scientists have been able to deduce carbon dioxide levels in the atmosphere over a period of 800,000 years. CO_2 levels were never above 300 parts per million until the year 1950. Since then they have risen consistently every year, and in 2019 they reached 416 parts per million. These are all critical issues that have to be understood and addressed, given the significant impacts they are already having on mankind. We must determine if any of these trends can be stopped or slowed and whether their impacts can be reversed. Burying our heads in the sand (or, in this case, in the polluted waters) is simply not an option. Another comment made by climate change deniers is that there are many scientists who don't believe in climate change. That too is simply untrue. A poll reported by NASA's Global Climate Change Initiative in late 2018 showed that 97.5% of climate scientists polled believe that global warming and climate change are real and can be attributed to human activities.

Let's look at some of these issues to get a sense of how they are being mitigated or dealt with (or not). The data establishing global warming as a fact are overwhelming. In the United States, the average temperature has increased between 2° and 3° Celsius since 1895, with the largest changes having occurred since 1970. Projecting forward, it is likely that temperatures will increase by an additional 1° to 2° over the next 20 or 30 years. These increases in terrestrial temperature are mirrored by evidence of changes in temperature high up in the atmosphere, as well as in the depths of the oceans, including the well-documented melting of glaciers and polar sea ice in both hemispheres. Furthermore, the increase in the frequency and severity of extreme weather events is well-documented. These include heat waves and cold waves, heavy rains, flooding, prolonged droughts and, of course, hurricanes and other intense storms happening around the world.

The impact of climate change on our health is far-reaching. Most obvious in urban settings is the sharp increase in asthma, a result of poor air quality associated with increased greenhouse gas emissions and temperature increases. But that's just one debilitating impact. We also see extreme heat conditions in different parts of the world that increasingly result in deaths from dehydration or cardiovascular events. Severe weather also results in injuries, fatalities and mental health issues caused by major storms and population displacement.

Global warming also results in changes in something called "vector ecology." In this sense, the vector is generally defined as the disease carrier, for instance the mosquito is the major carrier of the malaria parasite. Vector ecology refers to the vector's preferred habitat, which would then have an important impact on how and when the disease is transmitted. As temperatures and climates change, so does the range of insect-borne diseases such as malaria, dengue fever, Lyme disease, West Nile virus and a host of others. A number of viruses (and the insects that carry them) have moved north from equatorial areas of South America through Mexico and into the southern United States over a period of as little as 20 to 30 years. The correlation of these movements with changes in temperature and issues of water availability is very strong. What makes this even worse is that the diseases are moving from areas where the population has developed some tolerance to the infectious agents to areas where the population may be "naive" and therefore much more susceptible. We are seeing this repeatedly, as diseases, mostly viral, start to appear in North America when just a few years ago they were limited to the hotter, more equatorial areas of South America.

What holds true for insects can also hold true for humans. Severe environmental degradation — higher temperatures, absence of rain and groundwater resulting in less tillable land, and so on — often results in forced migration and civil conflicts. We've seen the impacts in large areas of North Africa and significant parts of the Asian continent. We also see, almost on a daily basis, the conflicts as groups attempt to escape the poverty and starvation in North Africa and migrate across the Mediterranean to southern Europe. These serious socioeconomic issues were brought about by a multitude of factors, but global warming and the growing inability of the people to feed themselves are most critical. Indeed, one of the most critical issues of climate change and global warming relates to the issue of food security, but I discussed that in the previous chapter, so here we'll just examine the impact of global warming on food production.

It may surprise you to learn that the impacts are not universally negative. In the more northern and southern latitudes, higher temperatures are bringing mostly beneficial effects. Higher temperatures in these areas may make more land suitable for farming and increase the length of the growing period, thus increasing yields. Higher temperatures may even increase yields in humid and temperate climates. In an almost bizarre twist, increased atmospheric carbon dioxide — a critical factor in global warming — may in some areas actually yield increases in crop production of as much as 20 percent (carbon dioxide being an essential component of photosynthesis and plant growth).

On the flip side, however, severe weather conditions — drought, flooding and extreme storms such as hurricanes — create havoc with our food supply. In recent years, drought conditions in Asia, sub-Saharan Africa, Australia and even parts of the United States and Canada have brought crop yields to zero in some areas. Conversely, flooding in some coastal areas, especially on the Asian continent (for example, India), has so contaminated farmland and rivers with saltwater that traditional farming becomes almost impossible.

Meanwhile, regulatory issues surrounding the contamination of lakes and rivers by overuse of fertilizers and subsequent runoff into the water supply are, together with the increasing cost of fertilizers, making the farming process ever-more complex and expensive.

The issue of global warming is intricately associated with a number of industrial, environmental and demographic problems that, in a sort of positive feedback loop, all tend to exacerbate the impact of global warming. Let's just look at a few examples.

What happens when the population increases? We need more food, so we need more land to grow food, so we cut down forests. Then there are fewer trees to use up the carbon dioxide, which results in increased atmospheric CO_2 (a greenhouse gas), a major factor in global warming. We also cut down forests because increased population often leads to increased urbanization, so even without burning fossil fuels as a source of energy — which is of course a critical problem — we are still increasing atmospheric CO_2.

With limited agricultural land available, increased food demand necessitates more productive growth, which requires increased use of nitrogen-based fertilizers. Whether they're mined (sodium nitrate) or produced chemically (ammonia), both types require a lot of energy, mostly as electricity, to produce. It's been estimated that fertilizer production uses as much as 50 percent of the total energy consumed by commercial agriculture. It may surprise you to learn that the amount of energy that goes into producing our various foods is very similar to the amount of energy our transportation industry uses. In other words, both contribute roughly equally to the problem of global warming.

And just in case you're thinking about it, the answer is not switching to all organic foods, as discussed in the previous chapter. Eating organic may be healthy, but it's likely to remain a Western luxury. Globally, we have far too many people to feed to even think about agriculture without man-made fertilizers to increase productivity. Someone once calculated that organic farming is so inefficient compared to modern commercial farming, that if the whole world were to convert to only eating organic foods, we would effectively have to decrease the world's population by well over 50 percent. So, eat organic food if you like, but don't mistake it for a solution to the world's food security or environmental crises.

Fossil Fuels

We know, of course, that the biggest cause of climate change is the burning of fossil fuels as a source of energy. If you think about that for a moment, it seems a bit crazy — surely there must be an easier way to fuel our lives. Under the current process, the sun provides energy to plants and the plants eventually die, but the energy remains stored in them in the form of carbon compounds. Those compounds are then mined, at great expense, millions or hundreds of millions of years later in the form of coal, natural gas and oil. The mining process itself is hugely energy-intensive; in other words, you have to use a lot of energy to extract energy from fossil fuels — extracting fossil fuels from the earth produces almost as much greenhouse gas as the burning of the fuel for energy. That just doesn't make sense; there must be easier ways.

The sun delivers more energy to Earth in an hour than we use in an entire year from fossil fuels, nuclear power and all renewable sources combined. Surely that should lead most of us to say, "There must be a better way." And there is. For decades, we've been looking for forms of energy that are cleaner, cheaper, safer and more renewable than fossil fuels. In addition to solar energy, renewable fossil fuel alternatives include tidal power, wave power and wind power. The advantages of these renewable forms of power are obvious. They are renewable, they are clean (in that they don't produce greenhouse gases or other pollutants) and they are present in many parts of the world. The disadvantages are that accumulating enough power from these sources to equal that which is generated from the burning of fossil fuels is almost impossible and, at least to date, both expensive and not especially reliable. We have not yet achieved the technology to store the energy from these sources, which are all to various degrees weather dependent. But perhaps most importantly, burning fossil fuels is just too easy. An important issue that is rarely publicly addressed is the cost differential between burning fossil fuels and the development of renewable or alternative sources of energy. Put bluntly, without appropriate incentives, it remains too easy (read too cheap) to just continue to burn traditional fuels and continue to pollute.

For many years, scientists and energy companies have examined the utility of burning biomass, such as garbage or forestry and agricultural waste, and of creating biofuels such as oil from corn or other oil-rich plants. The problem is that, for the most part, they have a similar carbon footprint to more traditional energy sources (petroleum products and electricity), at least in terms of their burning and their sources not being as reliable. These forms of carbon may be easier to source (they don't have to be mined), but their use as a fuel is the same. In the case of using corn oil as a biofuel, it has been suggested that this would cause increases in the price of both animal and human food and/or hunger as acreage is removed from food production and given over to energy production — not a very attractive trade-off. Assigning more acreage to growing such a biomass for fuel also requires us to cut down more forests, further exacerbating the greenhouse gas problem.

Synthetic Biology and the Fossil Fuel Problem

As synthetic biologists think about ways of addressing the fossil fuel problem, two immediate questions spring to mind: 1) Are there ways in which we can produce new fuels or utilize old fuels more efficiently? and 2) Are there ways in which we can lower levels of carbon dioxide and other greenhouse gases in the atmosphere with the intent to actually reverse global warming? Thankfully, the answer to both questions is yes.

Another important question, which has lingered for some time, is whether there are organisms that might take energy directly from the sun and convert that energy into oil or some form of biofuel without either going through the fossil phase, diminishing human or animal food supplies or requiring increased land use and therefore deforestation. One important and exciting answer lies in the slimy material we sometimes encounter on ponds and other still water — algae. Algae are very simple plant forms that may present as microscopic, in the form of pond scum (called microalgae) or as large seaweed structures (not surprisingly, called macroalgae).

The oil that microalgae produce is, at first blush, a superb biofuel. The oil is an essential component of the organism, as it is of all living cells. Energy produced from the oil is a largely renewable resource, since the algae extract carbon dioxide from the atmosphere as their own energy source and continue to grow as they produce the oil. The oil can be extracted, and the algae can continue to flourish. Since they are using carbon from the atmosphere as an energy source and basically recycling it into oil (the fuel), the process is virtually carbon neutral. Burning oil derived from microalgae produces a carbon footprint that is less than a tenth of that from an equivalent amount of coal or petroleum product or biofuel from corn, and that calculation only refers to the burning of the fuel, not its extraction. Furthermore, the algae grow quickly, doubling their mass within hours. Until recently, the downside to using microalgae-derived oil was that they typically grow on ponds and require lots of water and fertilizer, making the process fairly expensive. The size of the algae ponds and thus the costs required to produce enough oil to power a small village would be prohibitive.

Fortunately, a number of synthetic biology companies, including Craig Venter's Synthetic Genomics, are using engineering techniques to increase the productivity and cost-effectiveness of microalgae as a biofuel source. They are asking engineering questions with the goal of using the tools of synthetic biology to create solutions. These questions include:

- Can we modify, probably genetically, the algae to increase the amount of oil produced on a per-unit basis?
- Can we get the algae to grow more rapidly, thereby producing more oil faster?
- Can we grow the algae in a fluid compartment in a vertical format similar to sails, thereby increasing the area relative to growing them horizontally on the surface of ponds?

With respect to the rate at which the oil is produced, researchers at Synthetic Genomics have been able to double the amount produced by the algae on a per-unit basis. They've done this by genetically modifying a switch in the algae that converts the sugar made by photosynthesis, taking carbon dioxide from the atmosphere to increase the rate of conversion of carbon from sugar to oil, a more efficient fuel.

Encouraged by early successes such as this, major energy corporations are embracing the promise of synthetic biology and have made significant investments in improving these technologies to make the promise of algae biofuel a reality. As an example of how serious an initiative this is, as a bit of a paradox, ExxonMobil, one of the world's largest oil and gas companies, has invested in excess of $1 billion into Synthetic Genomics and the algae biofuel initiative.

SOLVING THE CARBON CONUNDRUM

It is abundantly clear that we have an accumulated carbon problem. Ever since the first industrial revolution, we have been emitting much more carbon dioxide into Earth's atmosphere than Earth is able to use for photosynthesis and plant growth or to deal with by simple absorption. The numbers are staggering. Humans have emitted 1.5 trillion tons of carbon dioxide into the atmosphere since the mid-19th century. That's three *quadrillion* or 3,000,000,000,000,000 pounds of carbon dioxide, and about half of that has remained in the atmosphere and is the major contributor to greenhouse gases and global warming. And to add insult to injury, while some of that carbon dioxide is effectively recycled and has been used for plant growth (e.g., photosynthesis), including forests, the appalling rates of deforestation worldwide only causes the carbon dioxide in the atmosphere to increase further. Atmospheric carbon dioxide and other greenhouse gases also equilibrate with and dissolve in our oceans and lakes, causing acidification, another serious form of pollution that has a devastating impact on aquatic life.

Inasmuch as the negative impacts of population growth and massive industrialization are man-made, so will the solutions have to be. The notion that these are just natural cycles and that nature can take care of herself is certainly passé. Some of the man-made solutions to our man-made problems will no doubt have to come from better stewardship of the environment: more efficient consumption of energy and water, decreased dependence on fossil fuels and increased use of renewable energy such as wind and solar power will be essential. Adoption will need to be encouraged through coercion — or, rather, incentives. A good example are the carbon taxes that are now being implemented. In support of the 2015 Paris Agreement, many of the world's governments have attempted to cut carbon emissions by establishing carbon taxes and tradable carbon credits and encouraging the establishment of carbon offsets. Unfortunately, even as this book goes to print there are governments that are reversing some of their commitment to the Paris Agreement in the misguided notion that either climate change is not as big a problem as some thought or, probably more likely, that the solutions are too expensive. It is hard to fathom how some might think that the strength of our economy and lower cost of goods are more important than reversing the damaging effects of global warming. That is short-term thinking that is surely going to come home to bite us.

Furthermore, much research is being undertaken in an attempt to find ways to "bury" carbon to ensure that it doesn't enter the atmosphere, where it would add to the greenhouse gas problem. Solutions such as storing the carbon in large caverns below the earth's surface or binding the carbon to solid compounds and then burying it are being actively pursued. While they are laudable in their acceptance of the critical implications of climate change and the role of carbon, these solutions seem futile at worst and short-term at best. It seems evident that we must figure out a way to stop the continued release of carbon into the atmosphere, and by taking big steps, not baby ones. And we must not only stop the emission and accumulation of greenhouse gases, we must also figure out a way to actually reduce the levels of CO_2 and other greenhouse gases in the atmosphere and in the oceans. But it is likely that those changes will come about too slowly, if for no other reasons than the preliminary costs will be too high, and many corporations and governments have horizons that are much too short. The Holy Grail of those who wish to reverse global warming is to not only reduce carbon emissions but also to recapture the excess carbon dioxide that has been released into the atmosphere. However, simply capturing and storing carbon will probably be only a stopgap measure, as there will be limits to how much carbon can be stored or hidden away.

The realization that increasing atmospheric carbon dioxide represents a major threat to society is now being well-recognized by major international corporations and venture capital companies that easily see potential profits in carbon capture, and that's a good thing, especially from those who have a long-term view. There are a number of exciting new initiatives in the pipeline . . . that's a terrible expression for this discussion, but you get my drift.

Climeworks, a Swiss company, has developed a series of physical plants that capture atmospheric carbon dioxide on filters, heat the carbon dioxide on the filters, release it from the filters and collect it as a concentrated carbon dioxide gas to supply to customers for a whole variety of industrial uses including producing building materials — such as cement — plastics, and fuels. It's almost as if you're capturing and reusing the carbon energy after burning fossil fuels. Depending on how much energy is required for the process (the company says that it's very little), this could be a very carbon-neutral process in terms of the use of the captured carbon as a fuel and a very effective carbon-lowering technology when the carbon is used in technologies that don't re-release it into the atmosphere.

In addition, over the past two to three years, a number of individuals and companies have begun using a variety of microbes and synthetic biology technologies for carbon capture and reuse of the carbon for value-added products.

A Harvard chemist, Daniel Nocera, is the inventor of the "bionic leaf." His original objective was to develop an artificial leaf to mimic photosynthesis and, in so doing, produce a cheap and plentiful source of energy using sunlight, water and carbon dioxide from air. He has recently reported on the engineering of a bacterium, *Ralstonia eutropha,* that efficiently converts carbon dioxide, hydrogen (from water) and oxygen into biomass and fuel alcohols. Nocera believes that the process is scalable and would not only provide a carbon-neutral fuel but that, at scale, could represent a major draw of carbon dioxide out of the atmosphere even to the point of reversing global warming.

Tobias Erb, a professor at the Max Planck Institute for Terrestrial Microbiology, is a prototype of a modern synthetic biologist. He is not into the business of tweaking living systems like microbes to coach or induce them into solving problems that we face in our environment. Rather, he uses the knowledge of biochemistry and genomics to build completely novel solutions. His approach uses known bacterial enzymes to pull carbon dioxide out of the atmosphere and convert it into useful carbon-based compounds such as biofuels or polymers. His technical advantage is that he uses an enzyme that is one hundred times faster than the enzymes that plants use in photosynthesis to extract carbon dioxide from the atmosphere. With this substantial numerical advantage, Erb is creating synthetic biology systems using engineered bacteria or plant cells to reduce greenhouse gases and produce economically important products.

And in California, LanzaTech, a now well-established synthetic biology company, starts with a simple supposition. Many of the products that we use every day in our routine lives have a carbon base. Typically we get that carbon from fossil fuels and that, of course, causes a problem, as carbon dioxide is released into the atmosphere and contributes to all the misery of global warming, to say nothing of the energy costs associated with mining the fossil fuels. LanzaTech wondered, why not instead get carbon from the air, thus killing two birds with one stone? (Sorry, another bad analogy.) If we could capture the airborne carbon, it would reduce the requirement for fossil fuels *and* reduce atmospheric carbon dioxide. The company's solution, just like many in synthetic biology, comes from microbes, in this case some of the earliest microbes known to man (at least four billion years old) — acetogens, or gas-fermenting organisms. These microbes grow by fermenting carbon-rich gases, as opposed to the much more common carbon sugar–fermenting microbes.

LanzaTech's microbes grow naturally on gas emissions from hydrothermal vents. Based on that background, scientists generate these microbes by using the carbon-rich gases emanating from a variety of industrial processes such as steel manufacturing, oil and gas production, agriculture, forestry and municipal waste. This effectively reduces emissions and uses the carbon to produce a variety of products, including biofuels, plastics and a host of other materials. Just a note of caution: plastics conjures up the image of plastic waste overwhelming our landfills and contaminating our oceans. But plastics are not bad per se; the problems lie in how we deal with their waste . . . or rather currently, how we don't. Nevertheless, the concept is beautifully simple. Rather than carbon capture and storage, which in the long term has to be futile, it's carbon capture and reuse — another simple lesson taught to us by billions of years of microbial biology. And if the carbon captured comes from the atmosphere and is used to make any number of products, then we are effectively *decreasing* carbon dioxide (the most serious of the greenhouse gases) in the atmosphere, and if we can do that at a sufficient scale, it should be possible to actually reverse global warming.

WATER

The quality and quantity of water available throughout many if not most parts of the world has become critical in terms of the very existence of many species. Once again, the issues are population growth, industrialization and how we shepherd our water resources. Think back for a moment to the Dust Bowl, the name given to the drought-stricken areas of the Southwestern United States in the early 1930s. Then multiply that many, many times over, as huge areas of dry plains areas in Africa, China, Russia, Australia and even the United States suffer from water shortages that create similar dust bowls and diminish agriculture or even bring it to a halt. That is what our future holds if we don't make meaningful changes.

A well-studied example involves the vast North China Plain where groundwater is rapidly decreasing as the water table becomes lower and lower. It was estimated that since 2002, the water table in that area had decreased by an average of six to eight billion tons of water *each* year, all attributed to dry weather and climate change plus over-exploitation of water resources by the ever-increasing needs of a rapidly growing population.

In many areas of North and South America and large areas of Western Europe, outright water shortages are seldom seen, but in other areas of the world the problems are often severe. In fact, in some regions, water may not be free and may barely exist, and we see daily tragedies of people starving for water as much as they starve for food. Most areas of the world (California and Israel being notable exceptions) have done a terrible job of managing their water resources, if they do anything at all. Deriving fresh water from our oceans through the process of desalination is very doable for those Western nations who can afford it. Some countries, like Israel, recycle huge percentages — up to 90 percent — of their "wastewater," whereas in other countries, such as Canada, water recycling is only a fraction of that. Just as many of us in the West have a hard time thinking about food shortages in other parts of the world, we also have difficulty fathoming the incredible scarcity of water in many areas of the world.

In addition, polluted waters carrying water-borne diseases such as typhoid, cholera, dysentery and many others result in the deaths of over 3.5 million people annually. But the pollution of our lands goes well beyond the "simple" fouling of our water and our air. One only has to turn to any number of pictures or documentaries of the Anthropocene (or google it) to see how land usage has changed the earth, how pristine forests have turned into concrete jungles, how landfills and toxic industrial waste sites pepper the countryside of developed industrialized countries. All of which has direct and indirect impacts on us in terms of exposure to toxic wastes, a decrease in the quality and quantity of our agricultural, lands and the deaths of forests and other green areas, among other impacts. Some lakes and rivers are so polluted by industrial, agricultural and human garbage that the water is not drinkable. Even the oceans are polluted with human waste, from plastic bags to oil spills. And think for a moment about the quality of the air we breathe, especially in our large cities — one third of all deaths from stroke, lung cancer and heart disease can be attributed to air pollution. Remember those photographs of Beijing during the Summer Olympics? While that might represent an extreme, there is no doubt that air quality in virtually all our urban settings represents a major challenge.

CLEANING UP AFTER OURSELVES

One of the most important solutions, of course, is to stop polluting. But as we know, that's not easy and, moreover, it's not sufficient. We also must figure out a way to clean up a lot of the mess that we've made. There are ways, and they are doable, and, as you may have guessed, in many instances we're going to have to take lessons from microbes and use the new technologies of synthetic biology.

Bioremediation

Bioremediation is defined as the use of naturally occurring or deliberately introduced microorganisms or other forms of life to break down and consume environmental pollutants; the objective, of course, being to clean up a polluted site.

You will likely recall the horrific *Deepwater Horizon* disaster in the Gulf of Mexico; it occurred on April 20, 2010, and the spilled oil wasn't contained until September 19 of that year. This was thought to be the largest oil spill the petroleum industry had ever dealt with. The impact was enormous: human lives were lost, and the environmental impact was expected to be vast — coastal wetlands destroyed, marine species devastated and an entire fishing industry, especially shrimping, likely destroyed or, at best, set back for decades. But five years later, much (but not all) had returned to normal.

The strong recovery can be attributed to two main factors. One was the effectiveness of the cleanup operation conducted by the company, British Petroleum, under the direction of the US government. The second, and probably much more important, cleanup process was nature's natural resilience, specifically the presence of oil- and gas-eating microbes on the ocean bed where the leak occurred. These naturally occurring microbes have adapted to grow around areas where gas and oil leak from the ocean floor. And they are not simply being good Samaritans; they actually use the oil and gas as their principal source of food and energy. Here we see again a natural evolution in microbial function.

But it's not always that simple, and humans' attempts to "impose" biodegradation have not always been successful. For example, biodegradation techniques were employed in Alaska after the *Exxon Valdez* spill in 1989 with very poor results (although these were early days in the development and use of these technologies). This was likely simply because the specific microbes that they used were unable to function in the cold waters off Alaska. Another failed attempt followed the *Amoco Cadiz* spill off the coast of Brittany in 1978. In that case, the cleanup process worked too well. The areas around the spill were effectively sterilized for a lengthy time, not so much by the oil spill but by overly aggressive efforts to contain the spill by using chemical leaching and bioremediation techniques. Clearly, what was required was a much greater deal of sophistication.

Today, using synthetic biology, we can expect the future of biodegradation using unique microbial populations to be much more effective. With our new sophisticated knowledge of microbial genomes and the specific degradative enzymes that they encode, plus the specific functional properties of the microbes — such as whether they work in cold temperatures, hot temperatures, fluids, soil and so forth — we should be able to design microbial populations to biodegrade specific types of oil spills or other contaminants. Now that we understand the process, we can use synthetic biology to create or modify existing "oil-eating" microbes to function in a whole range of different conditions, whether on land or in water, in freshwater or saltwater or in steamy or frigid conditions.

These technologies have moved well past the research lab and experimental use of microbial remediation in emergency oil spills. Today, environmental companies such as Tervita located in Calgary, Alberta, employs a host of different microbial solutions in its vast practice in the area of bioremediation with special attention to the "cleanup" of oil field contamination in Western Canada. Microbial

Discovery Group out of Wisconsin maintains a bacterial strain library of over two thousand different bacteria to attack a wide range of contaminating spills; including wastewater treatment, plant health, oil and other chemical spills. The approach of many of these companies seems to be that if you have a problem, we have or will develop a microbe to fix it.

Dealing with Mercury and Lead

We have been concerned for decades that industrial wastes have been adding toxic amounts of lead and mercury to our rivers and lakes. In many instances, this has led to the effective "killing" of lakes, as in nothing can grow in the polluted waters. Cleaning up these lakes is extremely difficult and very expensive. Fortunately, scientists have isolated microbes that thrive by effectively adsorbing, or "eating," mercury and lead in addition to other heavy metals. Various methodologies are being considered for using these bacteria to clean up our lakes and rivers. They include filtering contaminated waters through nets that have these heavy-metal-eating bacteria absorbed into them or using bacteria to precipitate the heavy metal (i.e., make it into a solid) and therefore stop its entry into the water system.

An understanding of the genomic adaptations that these bacteria have made in order to deal with substantial quantities of toxic heavy metals will help us design novel microbial systems to allow us to mine more cleanly. It is easy to imagine how similarly synthesized microbes could help keep other industrial sites from contaminating our environment.

PLASTIC JUNKIES

Without a doubt, many if not all of us have become plastic junkies (pun intended). Most of us are probably aware that the use of so much plasticware is bad for our environment, but we can't seem to kick the habit, and governments for the most part do not have the will to implement serious restrictions. Plastics, and specifically their disposal, represent one of the most critical environmental issues, leading to severe pollution of our lands, lakes, rivers and oceans, to say nothing of the energy used and CO_2 emitted in their manufacture.

The major component of most plastics is PET, or polyethylene terephthalate. In spite of the fact that it is an organic petroleum-based product, it does not degrade in landfills the way other organic refuse (such as wood, paper, grass and food scraps) does. It appears that while there are plenty of bacteria around to degrade those organic products, such is not the case for plastics, which are expected to remain in landfills for the long term. It's absolutely appalling to realize that normal plastics may take as long as a thousand years to decompose in typical landfills, plastic bottles could take 450 years, while our everyday plastic bags may take as "little" as 10 to 20 years. In open areas, including the oceans, degradation of plastics does occur, not as a result of biodegradation but, rather, caused by ultraviolet (UV) radiation from the sun. The results of such degradation are, however, unacceptable, since not only is it still quite slow, but the byproducts of PET plastic degradation are toxic to both sea life and humans.

Several attempts to produce biodegradable plastics have been made, with some modest success, but for a number of reasons they have not caught on. Efficiency and the cost of the production processes are often the major drawbacks, with the attitudes of the public and the lack of political will on the part of regulators also important factors. Simply putting a significant (not just a nickel) tax on all disposable plastic containers would encourage consumers to look for reusable containers and make it economically attractive to develop and introduce readily biodegradable plastic products.

Once again, an important solution likely lies in synthetic biology. In early 2016, a group of Japanese researchers identified a novel bacterium, labeled 201-F6, that appeared to be able to fully degrade plastic bottles, presumably by using the plastic as its primary source of carbon and energy. The uniqueness of these bacteria (as opposed to those found in most landfills) is that they have two enzymes that work together in tandem to completely break down the key ingredient of PET. Scientists are now identifying the two genes in the bacterium that code for these two enzymes. Utilizing the methodologies of synthetic biology, they will be able to produce microbes that break down plastics in conditions that range from landfills to recycling factories. One of the advantages of this particular process is that the breakdown products are both nontoxic and reusable.

SYNTHETIC BIOLOGY AND THE MINING INDUSTRY

Given the huge number of microbes in the earth, it is not surprising that they have an interesting potential interaction with the process of extracting minerals from the earth. There are a number of different prescribed interactions:

- **Biomining** is a technique of extracting specific metals from ores through the use of specific microbes. Bioleaching is essentially another term for biomining.
- **Bioaccumulation** is a process whereby microbes gradually accumulate substances, often toxic ones, from a specific site (air, water, earth, etc.).
- **Bioremediation**, as previously discussed, is the process whereby microbes specifically break down toxic or environmental pollutants, thereby cleaning up the contaminated site.

One of the most serious byproducts of both open-pit and closed-pit mining is the generation of large quantities of heavy metals (lead, mercury, cadmium and others) in the waste, or "tailings." While modest levels of these heavy metals are a requirement for life, including that of humans, higher concentrations are often highly toxic. Mine tailings contribute to contamination of groundwater and, eventually, our lakes and rivers.

However, many different bacteria have now been identified that have adapted to the presence of — and indeed thrive on — high concentrations of a wide variety of heavy metals. The BioZone in the Centre for Applied Biosciences and Bioengineering at the University of Toronto is committed to developing novel technologies, to remove heavy metals from a variety of contaminated sites by the process of bioaccumulation through genetically engineered microorganisms.

But their utility is not only in the process of bioremediation; bacteria can also be used to enhance the mining process at several different stages. In the process of "biomining" or "bioleaching," they can break the chemical bonds between copper and sulfate molecules or between gold and sulfide molecules, thereby releasing the copper or gold. This serves as a much gentler and more environmentally friendly approach to mineral extraction. This is not a new phenomenon; as early as 100 BC, Spanish copper miners would pour water from the Rio Tinto over deposits of copper sulfate, which would cause release of the copper. We now know that the red color of the Rio Tinto is a result of the presence of a specific bacterium that thrives on breaking the bond between copper and sulfate molecules, using the energy released from breaking the bond as its own food or energy source. Remember our earlier biology lesson — over billions of years of evolution, microbes have adapted to their environments. Here we described a microbe that adapted to use the energy from the copper sulfate bond as its own particular source of energy, and to the good fortune of the miner.

Other bacteria are adept at taking heavy metals and precipitating them as solids, ensuring that the toxic metals do not enter the aquifer and contaminate nearby lakes and rivers. A bacterium has been identified that converts soluble radium to an insoluble form, thereby ensuring that it doesn't leak out of mine tailings and create highly toxic and radioactive effluent, which could potentially pollute our aquifers and endanger our water supplies.

Universal Bio Mining is a Boston-based company coming out of MIT. It uses bacteria in the process of biomining to increase the returns in gold and copper mining (increasing the amount of ore successfully extracted) while reducing the environmental impact of using energy-dependent processes such as smelters and roasters, which are also the source of immense toxic output (CO_2 and SO_2).

As many of these valuable bacteria are sequenced and the genes/proteins associated with their adaptations to harsh and toxic situations are further understood, the importance and real potential of synthetic biology will be realized. It stands to become an essential component of humans' ability to live in a more harmonious fashion with the environment.

And just as I earlier described the potential production of entire new foods to grow and eat when we begin to live on Mars, we can similarly use the pertinent genes from many of these microbes to design new living organisms that can help us both clean up the environment on Earth and keep it clean. This might even represent a future positive phase of the Anthropocene epoch, one in which man is benefiting rather than harming Earth — and then maybe the attraction to go live on another planet like Mars and escape our increasingly polluted planet will not be so great.

Chapter Eight

IMPLEMENTATION

As we enter this new era, some will feel that we are moving into uncharted if not scary waters, but by now I hope you are as excited as I am by the hope and promise of synthetic biology. As we begin to consider how best to implement synthetic biology techniques into the mainstream, we will need to give careful consideration to education. By the nature of the many unknowns, new technologies almost always raise questions of ethics and regulation, and this will almost certainly be the case as synthetic biologists begin delivering solutions to many of our big three challenges.

This is going to be a fairly brief chapter, but not because I don't appreciate the critical importance of education, ethical behavior and adequate regulatory policies. Rather, it's because so many experts in the area are reflecting, opining and writing on these issues on an almost continuous basis. But I do have a few things to say, if for no other reason than I don't want you to think that any of these issues should be taken lightly. It's critically important that as we usher in this new era, we do so thoughtfully and prudently while at the same time recognizing that we are dealing with the fact that humanity is in crisis and thus timely development and consideration of these technologies will be essential.

I would be surprised and probably even a bit worried if some of you didn't find some of the concepts of synthetic biology a bit frightening. Learning from microbes, creating new life-forms (à la Craig Venter), moving away at least philosophically from natural selection (à la Charles Darwin) toward human-induced unnatural selection — these are very heady issues, and I'm pretty sure that we — scientists, academics, industry and government — have not done an adequate job of being fully transparent with the public about what types of developments are occurring, especially with regard to issues of safety and ethics. In this respect, scientists, administrators and regulators need to ask a number of critical questions in a public manner:

- Are we being transparent and acting in the public interest?
- Are technologies being pushed forward or held back for the appropriate reasons?
- Are ethicists and regulators equipped to make decisions about the issues they are charged to deal with?
- Are we dealing adequately with issues of risk, benefit and reward?
- Is politics getting in the way of progress?
- Are we giving too much power to special-interest groups (either for or against the implementation of a particular technology), allowing them to rule the day?
- Are we giving too much control to large corporations and profit-seekers?
- Are we dealing with the issues in a timely fashion?
- Are we doing an adequate job of educating the public?

These are all essential questions that should be considered, but at the same time, I don't believe we have the luxury of lots of time to pontificate. I don't agree with the approach of some who, when facing a difficult situation, would rather we stand back for a time and let the questions sit and fester and be discussed at academic forums ad infinitum. Why is that? It's because many of the issues we've discussed in earlier chapters are pressing. Humanity is in a very real crisis, especially as it pertains to the environment, where change and deterioration are occurring rapidly, and we have to find solutions rapidly. Without question, time is of the essence.

THE NAYSAYERS

While it is natural and healthy for there to be intelligent debate and discourse about new technologies, we must be certain that it is based on correct information. Lack of education and the spread of misinformation can lead to unwarranted, and fierce, opposition, as we see from groups such as anti-vaxxers, GMO opponents and climate change deniers. This anti-science trend has to be understood, countered and marginalized if we are to solve some of the big three challenges.

Because all of these synthetic biology solutions are based on sophisticated science that is often outside of the grasp of the average person, I would argue strenuously that governments, centers of learning and respected journals, magazines and newspapers all have to take the threats to our health, food supply and environment seriously as well as be open to potential solutions, including those that come from synthetic biology. Good, effective public education, at all levels, has to be paramount. To highlight how harmful a lack of information and subsequent opposition to science can be, below I discuss some of the most high-profile naysayers and the ramifications of the misinformation they continue to spread. Here I'm not talking about the crazies who would have us believe that the Earth is flat or that man never landed on the Moon. I'm talking about people who would do us serious harm as a result of their misguided ideology and/or their belief in "alternate facts."

The Anti-Vaxxers

At the same time as vaccine production is being revolutionized through the use of synthetic biology, a very strong anti-vaccine faction (the so-called anti-vaxxers) has taken hold, resulting in lower vaccination rates among children, which in turn results in the spread of certain diseases (e.g., measles) that we thought had just about been eradicated. While there have long been vaccine deniers, the utility of vaccination against a wide range of diseases simply cannot be refuted.

This most recent surge in anti-vaccine popularity is largely based on a 1998 article by British surgeon Andrew Wakefield. Published in a reputable journal, *The Lancet*, the article linked the MMR (measles, mumps and rubella) vaccine with the development of bowel disease and autism. Motivated by the fear of having their children develop autism, some parents began to refuse to have their children vaccinated, especially by the MMR vaccine.

In 2011, Wakefield's publication was found to be not only incorrect but actually fraudulent, and the British scientist was discredited, lost his license to practice medicine and was threatened with jail time. The findings were proven to be bogus, and the conclusion of the British Court was that "*there is now no respectable body of opinion that supports Wakefield's claims.*" However, by this time his false claims had already done untold damage to the public's opinion of the MMR vaccine specifically and to the general area of vaccination in general. To this day, Wakefield continues as an anti-vaccine crusader throughout the world and was recently hailed as a hero in Trump's new America. And to quote from one of Trump's notorious tweets: "Healthy young child goes to doctor, gets pumped with massive shot of many vaccines, doesn't feel good and changes — AUTISM. Many such cases!" WRONG! Check your facts Mr. President.

This is not trivial stuff. A critical problem with the anti-vax movement is that vaccinations are a public health issue and not merely an individual's preference. We understand that if we are vaccinated against a specific virus or bacteria, we are no longer likely to be infected or suffer the subsequent disease (e.g., measles, polio, smallpox). But anti-vaxxers ask: Why shouldn't I be able to refuse to be vaccinated or refuse to have my child vaccinated? Why is it anybody else's business? The reason is that when we don't get vaccinated, we put other members of our community at risk. The explanation for this lies in the concept of herd immunity, which may be defined as follows: "the resistance to the spread of a contagious disease within a population that results if a sufficiently high proportion of individuals are immune to the disease, especially through vaccination." In other words, the more people who are vaccinated, the lower the chance of the disease spreading. Therefore, if I refuse to be vaccinated, I am not only putting myself at risk, I am also putting both my neighbor and my community at risk.

The percentage of people who have to be vaccinated in order to achieve herd immunity hovers around 90 percent. With respect to the current outbreak of measles in specific regions of the Western world, the percentage of those vaccinated has moved from 90-plus percent 20 years ago to as little as 30 percent in some jurisdictions where, not surprisingly, there are major outbreaks of measles occurring and numerous resulting deaths. That all of this came about as a direct result of a fraudulent publication linking the MMR vaccine with the development of autism in young children and the promulgation of this linkage by a well-organized and very vocal group of individuals highlights the vital need for education and clear communication between scientists, governments and the public.

GMO Opponents

Similarly, there are a growing number of very vocal individuals, and even countries, who have developed overwhelmingly strong opinions that GMOs are dangerous. Even though there is not a shred of evidence linking any GMO to causing harm in any human or animal, the anti-GMO conspiracy continues to flourish and is evident not only on the grocery store shelves in Western countries, but also in agricultural policies in developing countries, especially in some African countries where GMO crops have been banned, much to the detriment of local farmers, citizens and their economies. The notion that GMOs are bad for our health has no basis in fact. It is utter nonsense, and yet it has become a major marketing tool of many large food corporations. What a sad commentary on our society. GMO opponents have been around for decades. Even though Norman Borlaug used only traditional crossbreeding in his work to produce improved wheat variants, some in the environmental movement criticized it as somehow unnatural and even to the detriment of subsistence farmers. But Borlaug was a visionary, and to such criticism he made a pointed retort that is even more critical today than when he said it in the early '80s:

> "Some of the environmental lobbyists of the Western nations are salt of the earth, but many of them are elitists. They've never experienced the physical sensation of hunger. They do their lobbying from comfortable office suites in Washington and Brussels. If they lived just one month amid the misery of the developing world, as I have for 50 years, they'd be crying out for tractors and fertilizer and irrigation canals and be outraged that fashionable elitists back home were trying to deny them these things."

Borlaug penned those words at least 60 years ago. I repeat the statement in its entirety because of its elegance and because it remains — and even more so, given the tenuous nature of the world's food supply — a pertinent response to those who continue to fight against the modern tools of biotechnology and, of course, GMOs.

Let me address an example of GMO resistance that has had devastating effects and is an important illustration of the dangers of misinformation. Vitamin A deficiency represents one of the worst diseases of childhood, resulting in the blindness and premature death of hundreds of thousands (750,000, at last count) of children per year from their lack of eyesight and from their compromised immune systems. In answer to the devastating impact of vitamin D deficiency, especially in developing countries, the Rockefeller Foundation began a search for a vitamin A–enriched rice (a GMO) in 1982. By 2002, Ingo Potrykus and Peter Beyer had genetically engineered a variety — called Golden Rice for its yellowish color — that incorporated a pathway for vitamin A. By 2010, clinical trials had made it clear that there were no safety concerns with respect to the genetically altered rice and that it was effective in reversing the effects of vitamin A deficiency, and in 2015 it was awarded a Patents for Humanity Award by the US Patent and Trademark Office. Yet today, children across the developing world continue to die as a result of vitamin A deficiency. And why is that?

Well, it's pretty simple. The Greenpeace organization and various other anti-GMO forces have worked effectively to make sure that Golden Rice is not available, both in the developed world and in the underdeveloped countries where children are suffering the most. Think about it for another minute: 750,000 children in underdeveloped countries die because a group of ill-informed elitists continues to spout anti-GMO nonsense, outright lies, really. I've tried to state the facts as I understand them, but in case that's not sufficient. A letter from 107 Nobel laureates in 2016 blasted Greenpeace for its opposition to GMOs and Golden Rice in particular. The letter in part reads:

> "Scientific and regulatory agencies around the world have repeatedly and consistently found crops and foods improved through biotechnology to be as safe as, if not safer than, those derived from any other method of production. There has never been a single confirmed case of a negative health outcome for humans or animals from their consumption. Their environmental impacts have been shown repeatedly

to be less damaging to the environment and a boon to global biodiversity.

Greenpeace has spearheaded the opposition to Golden Rice, which has the potential to reduce or eliminate much of the death and disease caused by vitamin A deficiency (VAD), which has the greatest impact on the poorest people in Africa and Southeast Asia."

A second example of the dangers of misinformation involves the impact of Monsanto's genetically engineered Bt corn on the monarch butterfly. Bt corn contains an inserted gene from a bacterium called *Bacillus thuringiensis*, which enables the corn to produce a biological toxin meant to kill pest insects, thus enhancing yields and allowing for a terrific decrease in the use of toxic chemical pesticides. What a wonderful idea and beneficial product. But a paper that appeared in *Nature* in 1999 concluded that pollen from Bt corn could kill monarch butterfly caterpillars and was thus the likely culprit in the observed decrease of monarch butterfly populations.

In 2001, the *Nature* study was discredited, and subsequently numerous investigations in the United States, by its Environmental Protection Agency, the National Academy of Sciences, the USDA's Agricultural Research Service and others, have all exonerated Bt corn. Rick Hellmich, an entomologist with the Agricultural Research Service, told *National Geographic* that:

"Butterflies are safer in a Bt cornfield than they are in a conventional cornfield, where they may be subjected to chemical pesticides that kill not just caterpillars but most insects in the field."

In spite of being discredited, the erroneous study led to the monarch butterfly becoming a sort of public symbol of the (falsely stated and misleading) environmental hazards of genetically modi-fied crops. To this day, that flawed study continues to fuel part of Greenpeace's fundraising and its aggressive, shameless but neverthe-less effective anti-GMO efforts.

The shortsighted opposition to GMOs and biotechnology-aided crops is putting major grain, especially corn, harvests in Africa at risk. While most corn grown in the Americas is the Bt maize that has been genetically altered to resist certain pests and therefore requires less pesticide use, African countries, with the recent exception of South Africa, have been religiously anti-GMO. This stance has been influenced and funded largely by European environmentalists, and we frequently hear nonsensical statements by ill-informed political leaders about the poisonous nature of GMO crops, sometimes even attributing it to an "American capitalist plot." This has tragic impli-cations. Unless attitudes change and the anti-GMO decisions are countered, African farmers will be forced to continue to use pesti-cides. They can usually ill afford these chemicals, which come with serious health, safety and environmental risks. What could follow, therefore, is a cycle of poor crop yields, increased famine and the necessity (where possible) of importing expensive foods.

One could easily argue that bans on GMOs, especially those that have been shown to be safe, are unethical. Several years before his death, a former director of Greenpeace said, "*I worry for Greenpeace and the other green groups because they could, by taking such a hard line . . . be seen to be putting ideology before the need for humanitar-ian action.*" In spite of this, the anti-GMO lobby continues unabated while people suffer and food supplies are at risk.

As an aside, and to show how non-scientific the basis of their stance is, it is interesting to note that the anti-GMO crowd seems to have given certain techniques of genetic modification a quite arbi-trary pass in terms of their GMO — and therefore "bad" — status. These include processes such as artificial selection for breeding and mutagenesis, or changes to the genetic information of an organism through either spontaneous mutation or as a result of exposure to a particular mutagen.

Climate Change Deniers

Finally, there are the climate change deniers. Even in the face of overwhelming evidence of climate change and, more specifically, the perils of the sharp increase in greenhouse gases and resulting global warming, deniers prevail and public efforts to curtail and reverse global warming are diminished. Many of them in conservative or right-wing governments across the Western world continue to put the short-term interests of their economies well before any attempts to reduce our dependence on burning fossil fuels and thus continue to pollute our environment in many different ways.

The science of global warming and the fact that it is directly linked to the growth of the world's population and increased industrialization is crystal clear. And yet in 2019, governments are retreating from the Paris Agreement based largely on false or short-term economic imperatives and canceling or curtailing carbon taxes that were meant to stem the tide of global warming.

Collectively these three groups of naysayers, and especially the anti-GMO and anti-vaccine movements, seem almost to be merging. Make no mistake, these are dangerous anti-science movements that are putting us all at risk. It is clearly all nonsense, but it serves to keep the average consumer confused and often scared. If you're interested in the issue of science literacy and the dangers of the anti-science movement, I would refer you to the Cornell Alliance for Science (www.allianceforscience.cornell.edu). Quoting from the Cornell group, "The most influential anti-GMO group in the US, the Organic Consumers Association (OCA), has also been directly involved in anti-vaccine campaigning. Earlier this year, OCA — alongside the anti-vaxxer groups the Vaccine Safety Council of Minnesota, the Minnesota Natural Health Coalition and the Minnesota Vaccine Freedom Coalition — organized a meeting targeting Somali-Americans in the state, among whom vaccination rates have plunged." This should not be tolerated.

So, who is to blame? It's easy to simply curse the misinformation or, in some cases, outright lies that are being espoused by the anti-vaxxer, anti-GMO and climate-change-denier forces. But, surely, we all share in the responsibility. Our regulatory processes and our public education campaigns should be communicating more effectively so that decisions are based on fact and not swayed by the forces of some particular agenda, political or otherwise.

I am obviously a strong advocate for modern genomics technologies and engineering (after all, that is the theme of this book). It's clear, however, that it would do no one any good to introduce new technologies that might cause harm, not only to ourselves but to all aspects of the environment. Furthermore, it would do untold harm to the continued development of this field. Together with governments, we scientists must ensure that we communicate the benefits and risks of any new technologies clearly and effectively to reduce the risk that misinformation or misunderstandings will inform potentially life-saving policies. For a much more in-depth discussion of the impact and implications of genomics, I point you to an excellent ongoing discussion of these issues directed by an initiative called the Genetic Literacy Project. They have a very effective motto. **"Science not Ideology."**

REGULATION

There is another issue that has to be addressed, and that is the timeliness and adequacy of public policies, especially including regulatory issues. The concept that "technology often leapfrogs policy" has always been concerning, but it seems especially so today, given the speed at which technologies are being developed and implemented, something that Thomas Friedman addresses so eloquently in his book *Thank You for Being Late.* We simply can't afford to allow the regulators to proceed at their too-often snail's pace. Regulators must adapt to the pace of change and be prepared to evaluate and make decisions in a timely fashion. Failing that, the public will be in danger; sophisticated technologies may start to enter common usage without due diligence and, therefore, without adequate protection for the consumer and ultimately society. It is essential that all new products of synthetic biology introduced into nature be subjected to rigorous regulatory and ethical standards. Hopefully those standards will also be both practical and timely.

Let me give you an example. Alphabet (Google's parent company) is bankrolling a start-up company, Verve Therapeutics, to see if they can use the CRISPR-Cas9 gene-editing system to develop DNA variations that confer on patients a resistance to the development of heart disease. One potential solution would be to introduce an edited gene to patients with high cholesterol, the edited gene being one that is known to enhance clearance of the so-called bad cholesterol by the liver. This would have the potential to permanently alter the patient's liver and, in turn, lower bad cholesterol in the blood, rather than to resorting to the usual cholesterol-lowering drugs that don't work exclusively in the liver and may have a number of mild to serious side effects.

This is an example of a sophisticated synthetic biology approach to a medical problem involving permanent alteration of a person's liver genome that could be extremely attractive to the patient and to the company producing the changed DNA. I'm not sure how good or how important an idea this particular solution is, but I do know that permanently altering an individual's genome is no trivial step; it needs careful but timely scrutiny. The regulatory bodies would have to be very diligent in dealing with the safety issues while at the same time being cognizant of the need and pressure to bring this potentially life-saving new treatment to the market. The regulators can't make any mistakes, but at the same time they have to recognize the public's need to get on with an exciting and potentially life-changing technology, especially in this era of rapid and exciting change. It's a tough balance, but it has to be achieved.

THE DEMOCRATIZATION OF SCIENCE

While earlier in the book, and this chapter in particular, I may have railed against "anti-science" forces, the fact is that science and technology are today moving forward at a more exciting and faster pace than ever. There has even been a movement of sorts, albeit still small at this stage, that might be termed the "democratization of science," driven by a number of different forces, including and even especially by the proponents and practitioners of synthetic biology (but not yet corporations) — and that's a good thing. For too many years science appeared to be primarily the responsibility of our senior academic institutions, our science-based industries and, to a lesser extent, certain government departments and agencies. For many years, it was men in white coats (science was largely male dominated), and they were for the most part rather isolated from the public. I know this from personal experience, as I would very often hear comments such as, "Wow, you're a scientist! I've never met a real live scientist."

But times have changed; science has become much more mainstream — and this is especially so with computer technologies and the ascent of the digital revolution. Our governments and educators are increasingly emphasizing STEM education, which stands for Science, Technology, Engineering and Mathematics. For many years, science-intense private-sector jobs were found mostly in a limited number of locations, such as Cambridge, Massachusetts; San Francisco and Silicon Valley in California; Cambridge, England and a few others. But now, in addition to the traditional science-based jobs in the engineering and pharmaceutical industries, many hundreds of thousands of people are working for companies like Microsoft, Apple, Google, Facebook and Amazon, and millions more work in small and medium-sized companies. You'll note that many of those new jobs are in information technology, but there is also a surge of job growth in smaller companies associated with biotechnology and the life sciences, not only in health but also in agriculture and in the environment. And the growth in synthetic biology initiatives has been astounding.

Perhaps the most exciting and certainly the most unexpected (in a good way) example of the democratization of science is the way in which young people have started to develop a strong and independent interest in synthetic biology. In January 2003, an independent study course at MIT challenged students to engineer biological devices to make cells — real live cells, usually microbes — blink. Yes, *blink*. Out of that early exposure to synthetic biology came an independent nonprofit organization "dedicated to education and competition, the advancement of synthetic biology and the development of an open community and collaboration." It was called the iGEM Foundation, which stands for International Genetically Engineered Machine Foundation, and its motto is "Synthetic Biology: Based on Standard Parts." Though it started as an independent study group, to quote from Arlo Guthrie's song "Alice's Restaurant" — "It became a movement." There are three components to iGEM's activities:

- a competition in which teams of students, mainly high school kids, are given a kit of biological parts and work over the summer to build and test novel ideas or products involving biological systems in living cells, ranging from bacteria to mammalian cells;
- a program by which academic labs can access the same resources that are made available in the student competition; and
- a registry of more than two thousand standard biological parts, which in many cases may be genes, with open access for all participating groups.

The growth of this program in the years since it was founded has been astounding, from a single lab at MIT in 2003 to more than 5,400 students in 325 teams from 35 different countries in 2018. The teams have developed hosts of interesting projects, some just fun and some quite serious, from rainbows of pigmented bacteria to banana- and wintergreen-scented bacteria to biosensors for the detection of toxic molecules possibly present in drinking water. Synthetic biology projects are now being developed for high school students to engineer new microbes at home with simple equipment, safe reagents and incredible, interesting outcomes.

In parallel with iGEM's services to young people, SynBioBeta is dedicated to nurturing stable growth in the now not-so-fledgling synthetic biology industry. The network holds an annual meeting, and it's interesting — although, in retrospect, not surprising — to note that the growth of synthetic biology has closely followed the decreases in costs to both sequence and synthesize DNA. Publications on the subject have expanded from only a handful in 2006 to well over three thousand in 2018, and investments in synthetic biology companies have increased dramatically, from less than $200 million in 2009 to over $3 billion in 2018.

Princeton scholar Susan Fiske is concerned (as am I) about the profile of science and scientists. She offers advice on how we can "Make Science Great Again" for the general public (please excuse any resemblance to presidential campaign claims). We start by understanding that, for the most part, at least historically, scientists are not seen as cool. Neither are they warm, and too often they are depicted as weird. Remember that ridiculous cartoon of Einstein with his white hair sticking widely on end? How could a kid possibly want to sit in front of a computer coding on a Friday night rather than go out to party with friends, or think that enjoying the "beautiful game" is watching a group of bacteria do battle with one another rather than taking in a soccer match? Fiske asserts the need for science to seem friendlier, for its (truer) image to be one of trying to make the world a better place, rather than the cold and distant reputation that science often has. She implores the scientific community to

- be transparent (explain what you do, even if it might be controversial),
- get involved (don't run away from communicating),
- educate (be a champion for STEM and practice education) and
- give purpose (you are trying to make the world a better place).

In many ways, synthetic biologists have taken Fiske's admonitions to heart. Using these parameters, they are attempting to move their image from distant, perhaps scary and sometimes even a bit worrisome or problematic, to a much more positive profile. They talk repeatedly of trying to find solutions for some of humanity's most critical issues and, following the lines in this book, in human health, food security and the environment.

But the kids in the iGEM programs are way ahead of us in capturing that "cool factor." Biohacker Nona Griffin equated her visit to an iGEM competition with a visit to Willy Wonka's chocolate factory, as in, **"Come with me, and you'll be in a world of pure imagination."** But it's about much more than simple imagination: the iGEM competitions are increasingly seeing these young people lead the way in a burgeoning field, using a variety of tools and a growing catalogue of gene parts to create a diverse and incredible group of novel organisms. While some of it may appear frivolous fun and games, much is driven by serious societal needs and a drive toward real utility and serious economic and commercial interest. Here are some examples:

- *Organofoam.* An iGEM team from Cornell University created an antifungal mechanism to prevent mold growing in biodegradable Styrofoam.

- *Detecting antibiotics in milk.* An iGEM team from the Beijing Institute of Technology built a detection system to test for the presence of antibiotics in milk — at home. The total project cost $30. That's 30 single dollars, not thousands or millions.

- *Biohacking blue genes.* An iGEM team from Berkeley, California, was able to produce a plant-based blue dye in bacteria. This is a potentially important replacement for indigo, an expensive and environmentally unfriendly oil-based dye that is currently used to color blue jeans.

- *Colisweeper.* A Zurich-based iGEM team created a new game somewhat equivalent to the digital game *Minesweeper*, one that uses live bacteria in a Petri dish and a pipet to make moves.

- *Phactory* A Munich-based team of young university graduates has developed a method to generate therapeutic bacteriophages that can identify and destroy harmful bacteria, especially those that are antimicrobial resistant. The potential importance of this cannot be overstated, given the scourge of AMR in many hospitals. It's most impressive that this was accomplished not by a high-profile university laboratory or a large pharmaceutical or biotechnology company but, rather, by a group of students putting their heads together and using some readily available technologies with minimal cost.

What's amazing about these projects is that not too many years ago, you could only imagine them being undertaken and, even then, only in sophisticated academic and biotechnology laboratories by professors and their PhD students and research associates and requiring tens or maybe hundreds of thousands of dollars and much more. Today, many of these projects have budgets in the hundreds of dollars or less and can be carried out at your kitchen table.

This is not just about high school "geeks" getting turned on by cool science and staying away from the soccer or baseball fields. Northwestern University in Chicago has developed a suite of educational kits (called BioBits) using the products of synthetic biology in simple solutions. The beauty of the system is that the students don't have to work with complex cells, which are often difficult to grow in culture and usually well beyond the technical capability of the average high school classroom. Instead the BioBits team removes the critical cellular machinery from inside the cells and freeze-dries it in order to create significant shelf stability. The students then carry out experiments with the material reconstituted into test tubes or even on paper. They're able to demonstrate gene editing using components of the CRISPR-Cas9. Using these cell-free systems, they were able to examine the role of the antibiotic resistance gene, and I quote from the work of Michael Jewett and Jessica Stark, the two principals in the establishment of BioBits Health at Northwestern: "In this module, students run two sets of reactions to produce a flowing fluorescent protein [this is their readout] — one set with an antibiotic resistance gene and one set without. Students then add antibiotics. If the experiment glows, then the fluorescent protein has been made, and the reaction has become resistant to antibiotics. If the experiment does not glow, then the antibiotic has worked." What's really neat about this reaction is that there is no danger of actually creating a drug or antibiotic resistance bacteria, because no live cells are being used, the entire process having taken place in a cell-free test tube. This is quite incredible. What an absolutely fantastic teaching tool; it allows high school students to see and do sophisticated biology and medicine in reactions previously available only to high-level graduate students in expensive research laboratories.

You might, of course, ask how all this is being controlled. Are there any regulations? The iGEM Foundation is way ahead of most of us in thinking about these issues, and especially in making contestants and organizers think about the issues. It forces them to consider the following questions:

- Why is this needed?
- Who will this help?
- Will this be abused?
- Who will control this?
- Who will pay for this?
- Will this cause harm?
- Who should be worried?

These are all important questions. The iGEM contestants are encouraged to think hard about the why and the how and to be able to defend the potential consequences of their work. This increasing, and well-considered, exposure to science is very exciting, as these young minds will no doubt play a key role in helping us solve some of our most critical problems.

FINAL THOUGHTS

In a very real sense, synthetic biology represents a new "field of dreams" — there seem to be no limits to what we can build. If there is a need, with these technologies, there will be a solution. The building or engineering of these new biological apps is without limit and, really, without precedent. With access to the genomes of millions of different organisms, from humans to the simplest single-cell microbes, and with the DNA codes and massive amounts of biology information accessible on our computers, we can now build or engineer any biological app that we can imagine.

So it's the year 2024 or maybe 2027 and here are some of the incredible things that we can do as a result of Synthetic Biology. We can be silly and build a carrot with two stalks that tastes like a tomato — no problem. We can increase the protein content of lettuce and grow it in brackish or salt-contaminated water. What a wonderful idea for poor areas of Asia where people suffer from malnourishment and from frequent coastal flooding of their farmlands. We can make rice that incorporates vitamin A to combat childhood blindness in developing countries — no, wait, that's already been done, and I've ranted about it earlier.

We can make vaccines very quickly on demand to immediately stamp out newly emerging infectious diseases. Think of it, we would have nipped the COVID-19 pandemic in the bud and avoided that world-wide shut down that occurred. We would have saved hundreds of thousands of lives to say nothing of the economy. We can correct genetic mutations in the brain cells of patients with schizophrenia . . . well, we can't quite do that yet, but the potential is at least clear.

We can make microbes that capture carbon dioxide from the atmosphere and reuse the carbon for a whole variety of products. This has the potential to reduce greenhouse gases and not simply stop global warming but actually reverse it. And we can use microbes to clean up mercury in our rivers and lakes.

The possibilities are without limit. If we know the biology and we have access to the genes, stored in our computers, we can build the solutions using synthetic biology and genomics. This is why I'm so optimistic. I believe in science and in young people developing solutions to many of our most critical problems related to our health, food supply and environment. And I'm convinced that the science of synthetic biology will be central in providing many of those solutions. So, today, as we enter this new era of synthetic biology, there is much more than just hope — there is promise.

Appendix A

A GENETICS PRIMER

In the following pages, we'll trace the development of genetics and genomics over the past 150 years — sort of a genetics primer. I'll offer a brief taste of the discoveries and especially the technologies that are now allowing us to address and even solve some of mankind's most serious problems in health, agriculture and the environment and to effectively shape the future. A glossary of terms follows the primer.

1856–63: Gregor Mendel studies inheritance traits in pea plants.

The origin of the term "Mendelian inheritance" stems from the name of this Augustinian friar who became known as the "father of modern genetics." People have known for millennia that offspring inherit traits from their parents, whether it's a child with her father's eye color or two plants crossbred to produce a new type. But it wasn't until Mendel that we started to understand this scientifically. Through his "Punnett square" studies with peas, Mendel showed that inherited characteristics were not simply a continuous blend of those of the parents but, rather, were distinct qualities passed on in a predictable way.

1902: Archibald Garrod discovers inborn errors of metabolism.

In the early 20th century, Mendel's work began to be used to understand human disease and inheritance. By observing families in which an obscure and poorly understood disease was present in both parents and children, Garrod came to realize that specific chemicals, present in urine or feces, represented a defect in metabolism, and that those defects (which were often fatal) could be passed down from parent to child.

1902: Boveri and Sutton prove that chromosomes are the units of inheritance.

Around the same time as Garrod's discoveries, new, more powerful microscopes facilitated the identification of the physical and chemical entities mentioned by Garrod. Theodor Boveri and Walter Sutton found that small objects within the nucleus of the cell, known as chromosomes, were what were physically passing down traits to daughter cells, though at the time, it still wasn't known what they were made up of.

1911: Morgan and Sturtevant demonstrate that genes reside on chromosomes.

Approximately 10 years after Boveri and Sutton's discovery of chromosome inheritance, Alfred Sturtevant and Thomas Hunt Morgan realized that chromosomes were made up of discrete pieces linked together, which were dubbed "genes" (the word gene can be traced back to the Greek word genos, meaning birth). These genes were, in fact, the distinct entities that Mendel and Garrod had postulated were responsible for inheritance.

1941: Tatum and Beadle discover that genes encode for proteins.

With their "one gene, one enzyme" hypothesis, Edward Tatum and George Beadle (along with others) pioneered molecular genetics by discovering that mutations in genes resulted in malfunctioning proteins in cells, proving that genes encode for proteins. This helped to explain how inherited genes result in diseases such as those studied by Garrod. It became the basis of what is known as the "central dogma of molecular biology," which still lies at the core of modern biology.

1943: Luria and Delbrück discover that genetic mutations can occur spontaneously in the absence of selection pressures.

Darwin's theory of evolution relies on random changes in organisms that provide an advantage to offspring. The experiment conducted by Nobel Prize–winners Max Delbrück and Salvador Luria showed that random genetic mutations can occur spontaneously, without any environmental stresses. Their research proved critical for our understanding of evolution.

1944: Avery, MacLeod and McCarty show that DNA is the material of genes.

Before the experiment carried out by Oswald Avery, Colin MacLeod and Maclyn McCarty, it was known that chromosomes were made up of protein and DNA, but it was unclear what encoded the genetic information. Their discovery proved that DNA was responsible and was essential for understanding inheritance at the molecular level, well before its actual structure was even known.

1952–53: Watson, Crick and Franklin discover the structure of DNA.

Rosalind Franklin's X-ray diffraction images of DNA allowed James Watson and Francis Crick to determine its three-dimensional structure. Arguably the most important discovery in biology, it gave us a literal picture of the famous double-helix structure of DNA, the basic unit of inheritance. Critically, it would usher in the era of genetic and genomic engineering that would change health, industry and agriculture forever.

1957: Kornberg produces the first in vitro synthesis of DNA.

Once we understood the structure of DNA, we could then treat it like any other chemical and create it in the lab. This was a major paradigm shift, upending the idea of the essentialism of living things that had existed since the time of Plato. Arthur Kornberg pioneered the first methods of making DNA from its chemical building blocks. If DNA was basically the "blueprint" of life, we now had the ability to make it in the laboratory. This is, of course, the basis of synthetic biology, and Kornberg could be thought of as its earliest father.

1961: Brenner, Jacob and Meselson discover the function of messenger RNA.

Following the discovery that RNA (ribonucleic acid) is found in the cytoplasm, the work of Sydney Brenner, François Jacob and Matthew Meselson demonstrated that messenger RNA is created from a DNA template. It serves to specify the amino acid sequence of proteins produced by a cell.

1961–67: Nirenberg and Matthaei crack the genetic code.

It was understood that DNA was a code made up of four base chemicals called nucleotides — A (adenine), C (cytosine), T (thymine) and G (guanine) — but it was unclear how the code was deciphered to make specific proteins, as per Tatum and Beadle's "one gene, one protein" theory. Marshall Nirenberg and his postdoctoral student, Heinrich Matthaei, cracked this code. They discovered which group of DNA bases coded for each of the 20 amino acids in the body, helping us to understand how genes lead to proteins. The central dogma of biology — that DNA leads to RNA, which leads to proteins — underpins all of our current understanding of disease and proper cell function.

1967–70: Alber and Meselson delineate the nature of bacterial restriction enzymes and DNA ligases.

This ushered in the era of us knowing how cells manipulate DNA. Natural enzymes were found in bacteria that "cut and paste" DNA. It was realized that they could be used as molecular "scissors" to selectively remove and insert genes, setting the stage for genetic engineering.

1972: Cohen and Boyer carry out the first recombinant DNA experiment.

Stanley Cohen and Herbert Boyer demonstrated for the first time that a gene from one species could be cut and pasted into another. This led to the first genetically modified organism — a mouse, in 1974 — synthetic human insulin in 1978 and the birth of the biotechnology industry. This revolution enabled the production of insulin in massive quantities (in fermenting vats, much like beer), replacing the previous practice of harvesting insulin from pigs.

1977: Sanger publishes the first complete DNA sequence.

Frederick Sanger and his colleagues developed the first DNA-sequencing technique, establishing the sequence of a simple virus and paving the way for more complex organisms. Once we could understand the genetic code, a push to read (i.e., establish the order of) the entire code of organisms — every A, C, T and G — began. A complete DNA sequence is known as a genome.

1983: The PCR technique is invented by Kary Mullis.

American biochemist Kary Mullis won the Nobel Prize for his use of the polymerase chain reaction (PCR), which allows for fast and easy replication of DNA. A key step in any biotechnology in use today is amplification (increasing quantity by replication) of DNA, whether it's to identify DNA from a crime scene or to use in DNA sequencing or genetic engineering. Mullis's technique is still the gold standard today.

1990: The Human Genome Project is launched.

One of the most audacious scientific endeavors ever undertaken (second, perhaps, only to the Moon landing) was launched with $3 billion in funding. Its intent was to sequence every A, C, T and G base in the human genome. The project was launched even before the technology was available to accomplish this goal, but its bold vision rallied the scientific community.

1994: The first transgenic crop is approved for sale.

The Flavr Savr tomato was the first genetically modified food approved for sale in the United States. Calgene developed the tomato by inserting a gene that prevented softening (caused by a natural enzyme in the tomato), which ultimately leads to rotting. The product was not commercially successful, but it paved the way for a market and for major new crops such as corn, soybeans and cotton to become almost entirely GMO.

1995–96: The first complete genome sequences are reported.

Technology developed by scientists involved in the Human Genome Project resulted in sequencing of baker's yeast and of the first bacterium. These were key accomplishments in their own right and also helped to develop technologies and pave the way for the human genome in the years to come.

2001: The human genome sequence is published.

When the Human Genome Project realized its goal, it capped off one of the greatest scientific feats in human history — on budget and earlier than planned. The importance of this accomplishment cannot be overstated. Genome sequencing changed how we study living things, allowing us to determine gene function and analyze complex systems of interactions responsible for cell functions and for disease. And, as you saw throughout the chapters of this book, it is the very basis of synthetic biology.

2001: The first fully synthetic virus is created.

Following on the Human Genome Project, a parallel path was initiated that aimed to synthesize genomes from scratch — from four bottles of A, C, T and G. This effort has been dubbed "synthetic biology," and it can be thought of as genetic engineering in the era of genomics.

2010: Thousands of human genomes have been sequenced.

Not only have human genomes been analyzed, so have the sequences of dozens of other species. A massive decrease in the cost allowed widespread sequencing of organisms, allowing us to catalogue genetic information unlike ever before. This created a rich data resource that can be mined to understand disease and develop new therapies.

2010: The first fully synthetic self-replicating cell is created.

J. Craig Venter, one of the leaders of the Human Genome Project, created a lab-made cell from four bottles of chemicals: A, C, T and G. Never before had a self-replicating organism been created from chemical building blocks — a significant technical and philosophical feat.

2012: CRISPR-Cas9 is repurposed to edit genomes.

CRISPR-Cas9, a bacterial immune system, was repurposed to simply and precisely edit genomes. The ability to edit DNA precisely, anywhere in the genome, has been revolutionary, speeding up basic biological research and creating gene therapies by which faulty genes such as those responsible for muscular dystrophy can be corrected inside the human body.

2016: Minimal synthetic bacteria are created.

Craig Venter took his synthetic cell work to another level by designing a genome from scratch, using a computer and printing the genome in the lab. The bacterium he created has the fewest genes of any living organism on Earth. It can be thought of as a blank canvas to which genes can be added to perform specific tasks such as synthesizing renewable chemicals, producing new vaccines — really, the possibilities are limited only by your imagination.

2016: The Human Genome Project–Write is proposed.

Following the successful synthesis of viral and bacterial genomes, George Church, Andrew Hessel and Jef Boeke proposed a new Human Genome Project: to synthesize the complete human genome, as well as those of other complex organisms like plants, from scratch. As with the original HGP, the tools and technology have not yet been developed to accomplish this, but it is envisioned that once more an audacious goal will catalyze the rapid development of technologies to usher in an era of engineering biology.

2018: Synthetic biology comes of age.

In 2018, over US$3 billion in venture capital was invested in 50 synthetic biology companies. They cover areas as diverse as therapeutics, aquaculture and biochemicals.

2019: Developments in genomics accelerate

On National DNA Day (April 25th) in 2018, the Broad Institute in Cambridge, Massachusetts, announced that it had sequenced one hundred thousand human genomes and the genome-sequencing company Veritas announced a cost to sequence of only $200, down from $300,000 just 20 years ago.

Appendix B

GLOSSARY

Acidification A result of uptake of carbon dioxide from the atmosphere by a body of water, resulting in a decrease in the water's pH (increased hydrogen means more acidity) that may be detrimental to marine life.

Algae Simple chlorophyll-containing organisms that live in water (salt or fresh). They range from single cells (often seen as scum on stillwater surfaces) to large multicellular structures such as seaweed.

Antibodies Proteins (also called immunoglobulins) generally produced by white cells to protect against a disease-causing substance such as a pathogen or bacterium.

Antigen A foreign molecule, sometimes toxic, that induces an immune reaction and the production of antibodies.

Antimicrobial An agent that kills microbes such as bacteria.

Antimicrobial resistance (AMR) A process whereby microbes become resistant to killing by antimicrobial agents.

Anthropocene A proposed epoch dating from the commencement of significant human impact on Earth's geology and ecosystem, including but not limited to anthropogenic climate change.

Anthropogenic Refers to something in nature that is caused or influenced by humans. Anthropogenic carbon dioxide is that proportion of carbon dioxide in the atmosphere that is produced directly by human activities.

Artificial cell Also known as a minimal cell, a genetically engineered particle that mimics some or all of the properties of a living cell and in every sense of the word is alive and can replicate itself.

Artificial intelligence Intellectual capacity demonstrated by a machine as opposed to a human or other living entity.

Bacterium A small single cell that is neither plant nor animal. Bacteria are tiny and generally live in communities of millions or billions in soil, water, air or the gut or skin. They make up at least 30 percent of Earth's total biomass. Most are harmless to humans, but a small number cause major diseases.

Biofuel A fuel produced through processes based on agriculture, forestry or biological waste as opposed to traditional fossil fuels.

Bioleaching Extraction of metals in the mining process with the aid of living organisms (usually microbes), often used to increase yields of purer minerals.

Biomining The process of using microorganisms to help extract metals of economic value in the mining process.

Bioremediation A waste-management technique that utilizes microorganisms (often bacteria) to neutralize and/or remove pollutants from contaminated sites.

Biotechnology The use of biological processes for industrial and other purposes, often referring to genetic manipulation of micro-organisms; for example, for the production of medical drugs or the creation of genetically modified (GMO) foods.

Brackish water Fresh water that has been contaminated with salt from the sea, usually unsuited for agriculture.

Bt corn A genetically modified form of corn (maize) that is resistant to certain crop-destroying pests. The first widely known GMO.

Carbon capture The process of capturing waste carbon dioxide (CO_2) and storing it to avoid producing more greenhouse gases.

Carbon dioxide A colorless gas made up of carbon and oxygen; the major product of the burning of fossil fuels and the main ingredient of greenhouse gases but also the main ingredient required for the growth of plants.

Carbon footprint The quantity of carbon dioxide or other carbon compounds emitted by consumption of fossil fuels as a source of energy.

Cellular agriculture The production of agricultural products from cell cultures as opposed to traditional field growth.

Climate change A change in the statistical (average) distribution of weather patterns when that change lasts for an extended period. Ice ages are a good example, as is the current period of global warming.

CRISPR-Cas9 A natural form of gene editing that certain bacteria use as a defense against attack by other microorganisms, particularly viruses.

DNA (deoxyribonucleic acid) A chain-like structure, consisting of molecules called nucleotides, that carries the genetic instructions used in growth, development, functioning and reproduction of all known living material.

DNA sequencing The process of determining the precise order of nucleotides within a DNA molecule, in which different sequences produce different instructions.

DNA synthesis The process whereby nucleotides are "stitched together" to form the chain structure of DNA, a process that occurs in almost all living cells but which can also be accomplished synthetically in the laboratory.

Evolution The change in inherited characteristics (genes) of biological populations over successive generations.

Food security The state of having reliable access to a sufficient quality of affordable, nutritious food.

GDP (gross domestic product) A standard measure of all the goods and services produced in a specific location over a period of time.

Gene editing A form of genetic engineering in which DNA is inserted, deleted, modified or replaced in the genome of a living organism.

Gene therapy Transplantation of a normal gene into a cell where a particular gene is either missing or defective.

Genetic engineering The direct manipulation of a cell's or organism's genes using the techniques of biotechnology.

Genetics The study of genes, genetic variation and heredity in living organisms.

Genome An organism's complete set of DNA, including all of its genes.

Genomics An interdisciplinary field of science that focuses on the structure, function, evolution and editing of the genome.

Gigabyte 1,000,000 million bytes (units of digital information), often used as a value for computer storage.

Global warming A gradual increase in the average surface temperature of Earth.

Global Land-Ocean Temp. Index The global temperature is the temperature measured just above the surface of the earth and averaged over thousands of measuring sites across the glove. The index establishes the period of 1950 to 1981 as the baseline set at 0.0 degrees Celsius.

GMO (genetically modified organism) A plant or food that has been modified to improve its quality and/or yields.

Golden Rice A genetically modified form of rice that through biotechnology incorporates a gene to produce beta-carotene, a precursor of vitamin A, hence its golden color.

Greenhouse gases Principally carbon dioxide, methane, nitrous oxide and fluorinated gases that trap heat in the atmosphere and therefore contribute to global warming.

Human Genome Project (HGP) An international effort, completed in 2001, to sequence the first entire human genome.

Hunger A severe shortage of food, resulting in serious malnourishment and possible death.

iGEM Foundation (International Genetically Engineered Machine) An organization that provides encouragement and support for synthetic biology to undergraduate students around the world.

Inflection point A point on a curve at which the curvature undergoes a substantial change.

Information technology The use of computers to store, retrieve, transmit or manipulate data or information.

Megahertz (MHz) One million hertz, a unit used to describe the speed of microprocessors, or how quickly computers process information.

Microbe A small bacteria, yeast or fungal cell; not including cells of either animal or plant origin.

Microbial genomics The study of the complete genome of a specific group or family of microbes, including evolutionary changes in specific microbes.

Natural selection The basic mechanism of evolution as defined by Darwin, referring to the differential survival and reproduction of single cells or multicellular organisms and resulting in the concept "survival of the fittest."

Nitrogen The most common element in the atmosphere (80 percent) and an essential component of amino acids, DNA and fertilizers.

Nonrandom mutation The purposeful (usually man-made) mutation of genes to effect a specific genetic modification. (See also *random mutation*.)

Nucleotide A specific molecular compound, consisting of a nucleoside linked to a phosphate, that forms the basic structural unit of nucleic acids such as DNA and hence genes.

Pathogen A microorganism that can cause disease, often a bacterium or virus.

Personalized/precision medicine An approach that tailors medical decisions, practices, interventions and/or products to individual patients based on their predicted response or risk of disease.

Population-based medicine Medical treatments based on evidence derived from averaging large groups of patients.

Random mutation A change in the DNA of a gene that occurs randomly or spontaneously, without apparent cause; a basic principle of evolutionary theory. (See also *nonrandom mutation*.)

Regulation Rules or laws established by government to protect the population.

RNA (ribonucleic acide) Like DNA, RNA is a chain-like structure consisting of molecules called nucleotides. One of its functions is to take the coded information from DNA and deliver it to sites in the cell where proteins are made.

Sequence The specific order or arrangement of nucleotides that make up DNA, and hence the genes that describe our genetic makeup. (See also *nucleotide*.)

STEM An acronym for science, technology, engineering and math, used to distinguish and encourage education in those fields.

Stem cell A cell in an organism that has not yet differentiated and can give rise to a number of different cell types (such as blood, liver or muscle).

SuperMeat An Israeli company dedicated to the use of cell culture techniques to make "real" meat from chicken or beef cells.

Sustainability The ability to sustain a status quo, as in the food supply or a specific ecosystem.

SynBioBeta A network of entrepreneurs, scientists and investors interested in and dedicated to the field of synthetic biology.

Synthetic biology The design and construction of new biological entities (enzymes, gene circuits or even whole cells) or the redesign of existing biological systems.

Synthetic genomics Arguably a synonym or subset of synthetic biology that deals with making specific genetic modifications to existing life-forms or producing entirely new gene circuits, new DNA and even completely novel life-forms.

Synthetic life-form An organism or novel "artificial" life-form derived entirely from synthetic (human-engineered) DNA.

Unnatural selection Intervention in the evolutionary process, usually by humans.

Urbanization A population shift from rural to city settings.

Vaccine A biological preparation (which can be an antigen or an antibody) that provides immunity to a particular disease.

Virus A small infectious agent that contains DNA plus a protein coat and which can replicate (divide) only within a living host cell.

Vitamin A deficiency (VAD) The absence of an adequate amount of vitamin A in the diet, a major cause of childhood deaths in underdeveloped countries.

World food supply The total amount of food available to sustain life on Earth.

Acknowledgments

First and foremost, I would like to acknowledge a friend and colleage Dennis McCormac, PhD, Associate Vice President, Science & Technology at Ontario Genomics. Dennis recognized the importance of synthetic biology at a time when most felt that it was a sort of academic curiosity, a bit of a fad and maybe even the stuff of science fiction. Dennis force-fed me article after article describing new ideas and advances in synthetic biology and showed me how critical these technologies are going to be to help solve some of the gravest problems that humanity faces. Thank you, Dennis, for all you do, especially for young people, and for being a champion for the beauty and importance of science.

Jordan Thomson, also from Ontario Genomics, was largely responsible for putting together the genetics primer found at the end of the book, a chronology of developments in genetics over the past 150 years. Tina McDivitt, Gayle Akler and Stuart Solway were instrumental in making me simplify some of the concepts to make them understandable to a lay audience. I hope we've accomplished that.

I also owe a tremendous amount to my mentors, some of whom I worked with over years on an almost daily basis and some of whom I followed closely but only knew by way of their work. Some of them are mentioned in the text. Tom Chang, Professor of

Physiology at McGill University, using the Socratic method taught me how to do science. Doug Wilson, former Dean of Medicine at the University of Alberta, taught me about leadership. Cal Stiller, who has worn innumerable hats in the Canadian medical science field, taught me about the business of science. And the amazing Lou Siminovitch, the effective Dean of Medical Research in Canada for over half a century, continuously instructed me on standards, integrity, how not to suffer fools and how to insist on excellence.

And finally, without an agent (the late Arnold Gosewich) and editors (Diana Byron & Lindsay Humphreys) and the folks at ECW Press (Jennifer Smith and Jack David), this work would not have come to fruition.

References and Additional Reading

CHAPTER 1

American Society of Human Genetics. "Six Things Everyone Should Know about Genetics." https://www.ashg.org/education/everyone_1.shtml.

Brown, Dan. *Origin*. Doubleday, 2017.

Church, George, and Ed Regis. *Regenesis: How Synthetic Biology Will Reinvent Nature and Ourselves*. Basic Books, 2012.

DoSomething.org. 11 Facts about Global Poverty. https://www.dosomething.org/us/facts/11-facts-about-global-poverty.

Engineering Biology Research Consortium (EBRC). https://www.ebrc.org.

Enriquez, Juan. *As the Future Catches You: How Genomics & Other Forces Are Changing Your Life, Work, Health & Wealth*. Three Rivers Press, 2001.

Friedman, Thomas. *Thank You for Being Late: An Optimist's Guide to Thriving in the Age of Accelerations*. Farrar, Straus and Giroux, 2016.

Isaacson, Walter. *Steve Jobs*. Simon & Schuster, 2011.

Koeze, Ella. "35 Years of American Death." Updated 2018. https://projects.fivethirtyeight.com/mortality-rates-united-states/.

Kurzweil, Ray. *The Age of Spiritual Machines*. Viking, 1999.

———. *How to Create a Mind: The Secret of Human Thought Revealed*. Viking, 2012.

National Human Genome Research Institute. Genetic Timeline. 2015. https://www.genome.gov/10506099/genetic-timeline/.

Norwegian University of Science and Technology, Department of Biotechnology and Food Science. "What Is Biotechnology?" https://www.ntnu.edu/ibt/about-us/what-is-biotechnology.

University of California, Berkeley, Center for Science, Technology, Medicine and Society. https://cstms.berkeley.edu.

U.S. National Library of Medicine. "What Are Genome Editing and CRISPR-Cas9?" Genetics Home Reference, 2019. https://ghr.nlm.nih.gov/primer/genomicresearch/genomeediting.

Wikipedia. "Genetics." en.wikipedia.org/wiki/Genetics

yourgenome.org. "What Is Genetic Engineering?" 2017. https://www.yourgenome.org/facts/what-is-genetic-engineering.

CHAPTER 2

Cepelewicz, Jordana. "Microbes May Rig Their DNA to Speed Up Evolution." *Wired*, August 19, 2017. https://www.wired.com/story/bacteria-may-rig-their-dna-to-speed-up-evolution/.

Enriquez, Juan, and Steve Gullans. *Evolving Ourselves: How Unnatural and Nonrandom Mutation Are Changing Life on Earth*. Portfolio/Penguin, 2015.

Grossman, Sara. "Evelyn M. Witkin." National Science and Technology Medals Foundation, 2002. https://www.nationalmedals.org/laureates/evelyn-m-witkin.

Hershey, Arlen. "Examples of Evolutionary Adaptation." Sciencing, April 19, 2018. https://sciencing.com/examples-evolutionary-adaptation-6131133.html.

Society for General Microbiology. *Microbiology Today* 31
(November 2004). https://microbiologysociety.org/publication/
past-issues/microbial-evolution.html. Issue dedicated to
microbial evolution.

Than, Ker. "What Is Darwin's Theory of Evolution?" Live
Science, February 26, 2018. https://www.livescience.com/474-
controversy-evolution-works.html.

CHAPTER 3

Friedman, Thomas. *Thank You for Being Late: An Optimist's
Guide to Thriving in the Age of Accelerations.* Farrar, Straus
and Giroux, 2016.

World Hunger Education Service. 2018 World Hunger and Poverty
Facts and Statistics. https://www.worldhunger.org/world-
hunger-and-poverty-facts-and-statistics/.

CHAPTER 4

Callaway, Ewen. "'Minimal' Cell Raises Stakes in Race to Harness
Synthetic Life." *Nature News* 531, no. 7596 (March 24, 2016).
https://www.nature.com/news/minimal-cell-raises-stakes-in-
race-to-harness-synthetic-life-1.19633.

Chien, T., A. Doshi and T. Danino. "Engineering Bacteria for
Cancer Therapy: Advances in Bacterial Cancer Therapies
Using Synthetic Biology." *Current Opinion in Systems Biology*
5 (October 2017): 1.

Church, George, and Ed Regis. *Regenesis: How Synthetic Biology
Will Reinvent Nature and Ourselves.* Basic Books, 2012.

Cornell Alliance for Science. www.allianceforscience.cornell.edu

Cumbers, John, and Karl Schmieder. *What's Your Bio Strategy?
How to Prepare Your Business for Gene Editing and Synthetic
Biology.* Pulp Bio Books, 2017.

Frost, Dan. "Startup Science: How the Idea for Synthetic Cells Took Silicon Valley by Storm." UCSF News Center, June 12, 2018. https://www.ucsf.edu/news/2018/06/410626/startup-science-how-wendell-lims-idea-synthetic-cells-took-silicon-valley-storm.

Hessel, Andrew. Programming Life. https://andrewhessel.com.

Hutchison, C.A., III, R.-Y. Chuang, V.N. Noskov, et al. "Design and Synthesis of a Minimal Bacterial Genome." *Science* 351, no. 6280 (March 25, 2016). http://science.sciencemag.org/content/351/6280/aad6253.

Interlandi, Jeneen. "The Church of George Church." *Popular Science*, May 27, 2015. https://www.popsci.com/church-george-church.

J. Craig Venter Institute. https://www.jcvi.org/.

McRae, Mike. "World's Smallest Tape Recorder Has Been Built Inside a Living Bacterium." ScienceAlert, November 24, 2017. www.sciencealert.com/smallest-tape-recorder-crispr-cas-bacterium.

Mullin, Emily. "CRISPR in 2018: Coming to a Human Near You." MIT Technology Review, December 18, 2017. https://www.technologyreview.com/s/609722/crispr-in-2018-coming-to-a-human-near-you/.

nature.com. Synthetic Biology: Latest Research and News. https://www.nature.com/subjects/synthetic-biology.

Ontario Genomics. "Think Synthetic Biology." http://www.ontariogenomics.ca/syntheticbiology/Ontario_Synthetic_Biology_Report_2016.pdf.

Pardee, Keith, A.A. Green, M.K. Takahashi, et al. "Rapid, Low-Cost Detection of Zika Virus Using Programmable Biomolecular Components." *Cell* 165, no. 5 (May 2016): 1255–66.

Rojahn, Susan Young. "Synthetic Biology Could Speed Flu Vaccine Production." *MIT Technology Review*, May 14, 2013. https://www.technologyreview.com/s/514661/synthetic-biology-could-speed-flu-vaccine-production/.

Skerrett, Patrick. "Experts Debate: Are We Playing with Fire When We Edit Human Genes?" *STAT News*, May 16, 2016. https://www.statnews.com/2015/11/17/gene-editing-embryo-crispr/.

Synthetic Biology Innovation Network. https://synbiobeta.com. "SynBioBeta is the premier activity hub for innovators, investors, engineers and thinkers who share a passion for using synthetic biology to build a better, more sustainable universe."

van Dongen, Maarten. "Synthetic 'Virus' to Kill Bacteria." AMR Insights, January 26, 2018. https://www.amr-insights.eu/synthetic-virus-to-kill-bacteria/.

CHAPTER 5

Centers for Disease Control and Prevention. National Center for Health Statistics. https://www.cdc.gov/nchs/index.htm.

Friedman, Thomas. *Thank You for Being Late: An Optimist's Guide to Thriving in the Age of Accelerations.* Farrar, Straus and Giroux, 2016. For growth curves of population, GDP, ocean acidity, Earth's temperature, atmospheric CO_2 and other factors versus time.

Lexchin, Joel. "The Pharmaceutical Industry in Contemporary Capitalism." *Monthly Review*, March 1, 2018.

National Cancer Institute. Cancer Statistics. https://www.cancer.gov/about-cancer/understanding/statistics.

———. "National Cancer Act of 1971." https://www.cancer.gov/about-nci/legislative/history/national-cancer-act-1971.

United States Department of Veterans Affairs. Multiple Sclerosis Centers of Excellence. https://www.va.gov/ms/.

Vigo, D., G. Thornicroft and R. Atun. "Estimating the True Burden of Mental Illness." *Lancet Psychiatry* 3 (2016): 171–78.

World Health Organization (WHO). Infectious Diseases. https://www.who.int/topics/infectious_diseases/en/.

———. World Health Statistics. http://apps.who.int/gho/data/node.main.1?lang=en. https://www.cdc.gov/nchs/index.htm.

CHAPTER 6

Brookes, G., and P. Barfoot. "Environmental Impacts of Genetically Modified (GM) Crop Use 1996–2015: Impacts on Pesticide Use and Carbon Emissions." *GM Crops & Food* 8, no. 2 (April 2017): 117–47.

Conserve Energy Future. https://www.conserve-energy-future.com. "Your source for green and sustainable living, environmental news and information."

Cribb, Julian. *The Global Food Crisis and What We Can Do to Avoid It*. University of California Press, 2010.

Food and Agriculture Organization (FAO). "World Food Summit." http://www.fao.org/WFS/.

Gao, C. "The Future of CRISPR Technologies in Agriculture." *Nature Reviews Molecular Cell Biology* 19, no. 5 (May 2018): 276–76.

Garthwaite, Josie. "Beyond GMOs: The Rise of Synthetic Biology." *The Atlantic*, September 25, 2014. https://www.theatlantic.com/technology/archive/2014/09/beyond-gmos-the-rise-of-synthetic-biology/380770/.

Golden Rice Project. "Vitamin A Deficiency–Related Disorders (VADD)." http://www.goldenrice.org/Content3-Why/why1_vad.php. "Haber–Bosch process." *Encyclopedia Britannica*.

McFadden, Johnjoe. "Synthetic Biology: The Best Hope for Mankind's Future." *The Guardian*, March 29, 2012.

Reynolds, Matt. "NASA Can't Send Humans to Mars Until It Gets the Food Right." *Wired UK*, March 26, 2018. https://www.wired.co.uk/article/food-in-space-mars-iss-station-astronaut-eating.

Roth, Elliot. "Solving World Hunger One Microbe at a Time." Spira, January 15, 2016. https://blog.drinkspira.com/solving-world-hunger-one-microbe-at-a-time- Smyth, Stuart. "25 Years of GMO Crops: Economic, Environmental and Human Health

Benefits." Genetic Literacy Project, April 6, 2018. https://geneticliteracyproject.org/2018/04/06/25-years-of-gmo-crops-economic-environmental-and-human-health-benefits/.

SuperMeat. https://www.supermeat.com.

Tierney, John. "Greens and Hunger; TierneyLab: Putting Ideas in Science to the Test." *New York Times*, May 19, 2008.

Umberger, Wendy, M. "Demographic Trends: Implications for Future Food Demand." Paper prepared for Agricultural Symposium, Federal Research Bank of Kansas City, July 14–15, 2015.

United States Department of Agriculture, VCell Productions and BeanCap. "Norman Borlaug and the Green Revolution." Uploaded January 27, 2012. https://www.youtube.com/watch?v=Lg9-HTtgFOk.

White-Stevens, Robert. *Pesticides in the Environment*. Marcel Dekker, 1971.

World Food Prize Foundation. "A Lifetime Fighting Hunger." Uploaded April 14, 2009. https://www.youtube.com/watch?v=m2TmEdiXTvc.

CHAPTER 7

Atlas, Ronald M., and Terry C. Hazen. "Biodegradation and Bioremediation: A Tale of the Two Worst Spills in US History." *Environmental Science & Technology* 45, no. 16 (2011): 6709–15.

Carson, Rachel. *Silent Spring*. Houghton-Mifflin, 1962.

Center for Sustainable Systems. "Carbon Footprint Factsheet" University of Michigan.

Ehrlich, Paul. *The Population Bomb*. Sierra Club/Ballantine, 1969.

Ginkgo Bioworks. "Biology by Design." https://www.ginkgobioworks.com.

Hylton, Wil S. "Craig Venter's Bugs Might Save the World." *New York Times*, May 30, 2012.

Karig, David K. "Cell-Free Synthetic Biology for Environmental Sensing and Remediation." *Current Opinion in Biotechnology* 45 (June 2017): 69–75.

National Aeronautics and Space Administration (NASA). "Global Climate Change: Vital Signs of the Planet." https://climate.nasa.gov/scientific-consensus/

———. "What's in a Name? Weather, Global Warming and Climate Change." https://climate.nasa.gov/resources/global-warming/.

Synthetic Genomics. "A San Diego biotechnology company that is harnessing the power of living cells — nature's most efficient machines — to create transformative medicine and bio-based products." https://www.syntheticgenomics.com.

Tang, Q., T. Lu and S-J. Liu. "Developing a Synthetic Biology Toolkit for *Comamonas testosteroni*, an Emerging Cellular Chassis for Bioremediation." *ACS Synthetic Biology* 7, no. 7 (June 2018): 1753–62.

Twain, Mark. "Notes on Innocents Abroad" [1904]. In *Autobiography of Mark Twain*, edited by B. Griffin, H.E. Smith, V. Fischer and M.B. Frank. Volume 2. University of California Press, 2010.

US Global Change Research Program. https://www.globalchange.gov/about.

World Health Organization (WHO). "Vector Control." https://www.who.int/vector-control/en/. Global approaches to vector-borne diseases.

Zhang, W., and D.R. Nielsen. "Synthetic Biology Applications in Industrial Microbiology." *Frontiers in Microbiology* 5 (August 2014): 451.

CHAPTER 8

Front Line Genomics Magazine, Issue 9 (August 2016). http://cdn. frontlinegenomics.com/wp-content/uploads/FG_Magazine_ Aug_2016_Full_Issue_V5-1.pdf. "A New World of Genomic Research."

Genetic Literacy Project. "Science, Not Ideology." https:// geneticliteracyproject.org.

Genome Canada. "Synthetic Biology: Biotech, the Next Generation." Press release, May 2, 2018. https://www.genomecanada.ca/en/ news/synthetic-biology-biotech-next-generation.

iGEM (International Genetically Engineered Machine) Foundation. www.igem.org. An independent nonprofit organization dedicated to the advancement of synthetic biology education and competitions and the development of an open community and collaboration.

Shanks, Pete. "When and How Will We Regulate Synthetic Biology?" Center for Genetics and Society, June 6, 2014. https:// www.geneticsandsociety.org/article/when-and-how-will-we- regulate-synthetic-biology.

Singh, Navyot. "Exploring the Disruptive Potential of Synthetic Biology." Interview with David Berry, Andras Forgacs and Ellen Jorgensen. McKinsey & Company, June 2016. https://www. mckinsey.com/industries/pharmaceuticals-and-medical-products/ our-insights/exploring-the-disruptive-potential-of-synthetic- biology.

SynbiCITE. http://www.synbicite.com. A UK synthetic biology and engineering biology industrial accelerator dedicated to commercializing the use of technologies such as CRISPR, gene editing and gene drives.

Index

demyelinating disease, 69
dengue fever, 124
depression, 4–5, 68, 71, 75
developed world. *See* industrialized countries
developing countries
  acceptance of GMO food, 98
  agricultural policies related to GMO, 148
  demand for protein, 113
digital revolution, 3, 7
diptheria, 52, 63
disability, 71
disease, 10, 62, 86
disease eradication, 52
disease prevention through vaccination, 61
disease resistant wheat, 91
DNA, 13, 22, 72
  "blueprint" of life, 5, 168
  changes in or mutations, 11
  chemical components of, 110, 169–71
  costs to sequence and synthesize, 41, 158
  data captured in, 59–60
  discovery of the structure of, 38, 43
  double-helix structure of, 168
  material of genes, 167
DNA-repair enzyme, 50
DNA-repair process, 15
DNA sequencing, 39, 41, 110, 158, 170
  inexpensive sequence to function at the bedside, 83
DNA synthesis, 39, 44, 110
DNA variations, 155
drip irrigation, 102
droughts, 4, 86, 100, 124–25

drug therapy in treatment of mental disease, 71
Duchenne muscular dystrophy (DMD), 74
Dust Bowl, 135
dysentery, 52, 136
dystrophin gene, 74

"Earth system," 27–28
Earth's temperature, 100
Ebola virus, 3, 5, 47, 62–64
Editas Medicine, 57
"educated" consumer, 98
educating public, 146, 148, 154. *See also* democratization of science
educating public on synthetic biology, 145
educational kits using products of synthetic biology, 161
Ehrlich, Paul, 91
  The Population Bomb, 90
electrical engineering, 38
electricity, 89, 126
electroconvulsive therapy, 71
engineering biology, 173
Engineering Biology Research Consortium, 5
Enriquez, Juan, 20–21
  As the Future Catches You, 46
  Evolving Ourselves, 19, 47
environment, ix, 1, 3–4, 30, 35, 118
  biotechnology in, 157
  changes to, 94
  crisis of, 146
  pressures on, 92, 119–20
environmental degradation, 125
environmental impact study of GM crops, 99
environmental issues, 23, 93

microbes cont'd
  genetically modified to help
    remove mercury and lead, 2,
    139, 163
  harnessing the powers of, 17–19
  learning from, 145
  replacing mined or chemically
    synthesized fertilizer with, 115
  techniques to sequester
    contaminants, 6
  that break down plastics, 141
  wonders of, 14–17
microbial biology, 135
Microbial Discovery Group, 138–39
microbial engineering and machine
  learning, 106
microbial evolution, 11
microbial genomes, 138
microbial genomics, 38
microbial research, 14
microbiology, 38
Microsoft, 45, 156
microsurgery, 53
middle class, growth of, 87
milk, 114, 160
milk without cows, 108
Minesweeper (game), 160
mining industry, 120
  contaminated mining sites, 119
  mine tailings, 23, 141–42
  open-pit and closed-pit mining,
    141
  synthetic biology and, 141–43
Minnesota Vaccine Freedom
  Coalition, 154
MMR (measles, mumps and
  rubella) vaccine, 147–48
molecular biology, 38, 55
molecular engineering, 38
molecular genetics, 50

molecular scissors (enzymes), 21–22
molecular tape recorder, 59
monarch butterfly, 10, 151
monoculture farming, 116
Monsanto's genetically engineered
  Bt corn, 151
mood stabilizers, 71
Moore's law, 41
Morgan, Thomas Hunt, 166
mosquito, 56, 124
Mullis, Kary, 170
multiple sclerosis (MS), 68–70
mumps, 61, 63
muscular dystrophy, 21, 55, 57,
  73–74
  Duchenne muscular dystrophy
    (DMD), 74
mutation, ix, 11, 55, 76, 82
  nonrandom mutation, 20–21

nanomaterials, 112
NASA's Global Climate Change
  Initiative, 123
NASA's Jet Propulsion Laboratory,
  120
National Academy of Sciences, 151
National DNA Day, 173
National Institutes of Health, 45, 79
natural foods, 98
"natural gene therapy," 73
natural selection (Darwin's theory
  of), 11–12, 19–21, 145
"natural technology," 22
neurological diseases, 53, 60, 68–70
new foods and new food groups,
  103–4
new foods to suit our needs, 109–10
Nigeria, 86
Nirenberg, Marshall, 169
nitrates, 110, 114, 119